ENVIRONMENT SCOTLAND: PROSPECTS FOR SUSTAINABILITY

Environment Scotland: Prospects for Sustainability

Edited by

ELEANOR McDOWELL
Glasgow Caledonian University

JAMES McCORMICK
Scottish Council Foundation

ROUTLEDGE
Routledge
Taylor & Francis Group
LONDON AND NEW YORK

First published 1999 by Ashgate Publishing

Reissued 2018 by Routledge
4 Park Square, Milton Park, Abingdon, Oxon, OX14 4RN
605 Third Avenue, New York, NY 10158

Routledge is an imprint of the Taylor & Francis Group, an informa business

A Library of Congress record exists under LC control number: 98074135

ISBN 13: 978-1-138-31312-5 (hbk)
ISBN 13: 978-0-429-45775-3 (ebk)

Contents

List of Contributors

Eleanor McDowell is a Lecturer in the Department of Social Sciences at Glasgow Caledonian University. Her teaching and research interests include environmental sociology, sustainable development, and environmental participation. Her PhD from Strathclyde University (1993) focused on the Scottish Environmental Movement. Recent publications include *The Historical Development of the Scottish Environmental Movement*, Caledonian Papers (1996) and 'Reconnecting Communities' (with Douglas Chalmers) in *Learning to Sustain* (ed. J. C. Smyth, 1998). She is currently chair of the Scottish Environmental Research Group at GCU.

James McCormick is Research Director of the Scottish Council Foundation. Prior to that he was a Research Fellow at the Institute for Public Policy Research (IPPR) in London, where he worked for the Commission on Social Justice, related welfare projects and on Scottish devolution. Recent publications include *Welfare in Working Order* (with Carey Oppenheim, IPPR), *Paying for Peace of Mind* (IPPR/PSI) and *Three Nations: Social Exclusion in Scotland* (SCF).

David Begg is Professor of Transport at Robert Gordon University and Convener of Transportation on the City of Edinburgh Council. In addition he advises government at Westminster and local levels as a member of the Transport White Paper Advisers Team and Transport Spokesperson for the Convention of Scottish Local Authorities (COSLA). He has been a local government councillor in Edinburgh since 1986.

Lynn Bennie is a Lecturer in Politics in the University of Aberdeen's Department of Politics and International Relations. Her research interests are in green politics, Scottish politics and Britain's political parties. Recent publications include (with J. Mitchell and J. Brand) *How Scotland Votes: Scottish Parties and Elections*, Manchester University Press 1997. She is currently completing a PhD thesis on the Scottish Green Party.

Sarah Boyack is a Lecturer at Edinburgh College of Art (School of Housing and Planning). Her areas of interest include the operation of the planning system, particularly development plans and strategic planning. She chaired the John Wheatley Centre's Environment Commission, which published *Working for Sustainability: An Environmental Agenda for a Scottish Parliament* in August 1997. She was appointed to the Secretary of State's National Transport Forum for Scotland in 1997. She was part of the research team which published the *Review of Development Planning in Scotland* for the Scottish Office Development Department in March 1998. She is the Labour Party's Scottish Parliament candidate for Edinburgh Central.

Douglas Chalmers is a Postgraduate Student in the Department of Economics at Glasgow Caledonian University. He is currently writing a PhD on Language, Culture and Economic Development. His current teaching and research interests include environmental and sustainable economics, sustainable development and economic history. Recent publications include 'Reconnecting Communities' (with Eleanor McDowell) in *Learning to Sustain* (ed. J. C. Smyth, 1998).

Alistair Dalziel is a Director with the RPS Group PLC with responsibility for the West of Scotland Office. He is a qualified chemist, holds an MBA and an accountancy qualification. He was previously a Director of Tyne & Wear Development Corporation with responsibility for regeneration of the riverside areas of Newcastle and Sunderland.

Kevin Dunion has been Director of Friends of the Earth Scotland since 1991 and Chairman of Friends of the Earth International since 1996. He is a member of the Secretary of State for Scotland's Advisory Group on Sustainable Development. Publications include *Living in the Real World - The International Role for Scotland's Parliament* (1995). He regularly represents FoE International at meetings with the OECD, the UN Commission on Sustainable Development, European Environment Ministers, and the European Commission.

Tony Gloyne is a Lecturer in Glasgow's three University Economics Departments, with additional involvement in Engineering Economics and Environmental Economics. He has undertaken extensive energy-related consultancy work on behalf of the European Commission, the UK Department of Energy, the UN Food & Agriculture Organisation (FAO) and Euratom. He is a visiting Professor at the University of Zulia, Venezuela (with the support of the British Council).

Alan Hutton is a Senior Lecturer in Economics at Glasgow Caledonian University. He teaches and researches on the Scottish tradition of political economy. He has a particular interest in the interaction of government and private firms in the pursuit of the public interest. His interest in the environmental aspects of energy use arises from an involvement in the regulation process as a member (since 1995) of the Southern Scotland Electricity Consumers' Committee.

Andrew Raven is Director of Land Management of the John Muir Trust, the charity dedicated to conserving wild places for nature and people. He has wide experience of rural environmental, community and land use issues. A rural practice Chartered Surveyor, he is also Vice Chairman of Rural Forum Scotland, a Trustee of the Millennium Forest for Scotland, a member of the Scottish Consumer Council and of the Governing Body of the Macaulay Land Use Research Institute, and a Director of Ardtornish Estate, Morvern. He writes here in a personal capacity.

Hugh Raven specialises in environmental and international development issues. He is Convenor of the Green Globe Task Force, the advisory group to the Foreign Secretary and Environment Minister on international environment policy. He is a consultant on food, environment and agriculture policy. From 1990 to 1995 he was co-ordinator of the Sustainable Agriculture, Food and Environment (SAFE) Alliance. He chairs the Lochaber Fisheries Trust and is a trustee of the Soil Association and the RSPB. He divides his time between London and the West Highlands where he and his wife are restoring a 500-year old ruin, and he is a director of a farming and land management company. He is the Labour Party's Scottish Parliament candidate for Argyll and Bute.

David Silbergh is a Lecturer in Research Methods in the School of Public Administration and Law at the Robert Gordon University. He has broad research interests relating to public policy for sustainability and sustainable development especially with respect to rural Scotland.

Peter Strachan is a Director at the Centre for Industry and the Environment, based at the Robert Gordon University, Aberdeen. The purpose of this Centre is to explore the interface between organisational learning, innovation and environmental performance in business and public organisations. Peter has published in a number of journals, specialising in change management and organisational development and has recently published a best selling book *Managing Green Teams: Environmental Change in Organisations and Networks* (Greenleaf Publishing 1998). Peter also has much experience of management consultancy and currently works with a multinational corporation in Germany and Belgium.

Acknowledgements

We would like to thank all participants in the *Environment Scotland* seminar held in Glasgow on 5-6 September 1997. This collection is the direct result of the seminar and the discussion, redrafts and debate that followed. We benefited enormously from the inter-disciplinary nature of the seminar. Much of the creative thinking is reflected in the authors' contributions. We are also indebted to Cathy McCormack and Foster Evans, who spared the time to be interviewed about their involvement in their respective grassroots campaigns. We would like to thank Chris McWilliams for providing helpful comments on an earlier draft of the text. We would also like to thank Kirsty MacDonald of Market Research Scotland for her assistance with the company's Scottish Omnibus Survey. Colleagues at the Institute for Public Policy Research (IPPR) offered time and advice in the formative stages of this project. Finally both editors wish to gratefully acknowledge the research grant received from the Department of Social Sciences at Glasgow Caledonian University. This funding paved the way for our seminar to take place and for the subsequent public opinion survey.

Eleanor McDowell
James McCormick

1 Environment Scotland: An Overview

ELEANOR McDOWELL and JAMES McCORMICK

The idea for *Environment Scotland: Prospects for Sustainability* came from a discussion about the need for more detailed thinking on environment policy in Scotland. A group of policy experts met in Glasgow on the eve of the devolution referendum for a two-day seminar to explore: how environment policy might change; how it ought to change; what historical experience and international comparisons have to teach us; and the limits to achieving sustainability in a small country. The focus is deliberately on only the environmental dimension of sustainable development. Our focus could just as appropriately have been on how the labour market, public health or education contribute to (or detract from) the same goals.

This collection of essays presents the findings of the seminar and subsequent thinking. It brings together academics with business practitioners and voluntary sector experts, to explore a series of short and long-term challenges facing Scotland. The book benefits from the in-depth discussion shared by the authors and others who contributed to the Glasgow seminar. But it is not written in the style of a report by a commission. It does not offer a definitive set of recommendations supported by each author, although there is considerable agreement among them on what needs to change. Reflecting the diverse backgrounds of the authors, we expect this book will appeal to a wide audience spanning environmental studies, Scottish affairs, social policy and local government studies.

Introduction

From the late 1960s to the mid-1970s, environmental issues were synonymous with apocalyptic warnings of population explosion, pollution and depletion of the earth's natural resources at an unsustainable rate. They defined what one contributor to this book describes as the 'Armageddon Now' tendency. Demands for urgent action to avoid the worst-case scenarios were published through various blueprints, conference events and the formation of environmental groups and organisations. By the late 1980s, the environment had become popularised, politicised and commodified.

'Sustainable development' has become the green narrative for the 1990s. It offers a higher quality of life while respecting the principle of justice between the generations. It was the driving force behind the 1992 United Nations Conference on Environment and Development (UNCED) in Rio de Janeiro - popularly known as the Earth Summit. Attention focused on climate change, biodiversity and forest management. Perhaps the most influential outcome was Agenda 21, promoting local involvement in planning and implementing measures towards sustainable development. 'Think local - Act global ' became the popular term for expressing the interconnectedness of individual and community action with wider environmental objectives.

Earth Summit 2 took place five years later in New York. It reviewed the commitments made in Rio and established a series of priorities for the future. Both Earth Summit events raised public awareness and expectations. Nevertheless, commentators have called for greater political commitment on the part of national governments and voiced concern over the deepening inequalities in well-being and environmental standards between industrialised and non-industrialised countries. A five-year review of the 1995 Copenhagen World Summit for Social Development is planned for 2000. In preparation for the Third Earth Summit in 2002, poverty and the links between human and environmental degradation will be the leading concerns for governments and the unofficial groups who meet on the fringes of such events. Sustainable development has provided a focus to unite disparate interests in dialogue if not action. The challenges are more daunting than ever: that much is uncontroversial. The extent of the problems to be solved threaten to defeat even the most committed environmentalist, and obscure some of the first principles of sustainable development. Most of the problems can only be addressed by international action and are beyond the focus of a book on what can be achieved in Scotland. Some have been created by decision-makers who are elected and can expect to be held to account for their actions; others may wield more economic power without accountability being clearly located.

While there is growing interest in the implications of sustainable development in Scotland, the distinctive Scottish features of sustainability have received insufficient attention from academics, practitioners and policy-makers. This collection of essays seeks to improve on that position. Ahead of the establishment of a devolved Scottish Parliament, this book asks:

- What will be the key policy options and constraints around sustainability?
- What practical difference will devolution make?
- How will policy-making across the different tiers of government adjust to the demands of sustainability?
- What changes beyond government will be required to move beyond the rhetoric of sustainable development?

The chapters are organised into four sections. This chapter presents a summary of the key theoretical and practical issues raised in each contribution, indicating where the lines of agreement and disagreement lie. It is followed by a comprehensive review of the theoretical foundations underpinning 'sustainable development'. It is not an easy concept. Like 'partnership', 'social exclusion' and perhaps 'stakeholding,' it is difficult to define and even harder to achieve. In Chapter Two, David Silbergh traces how the concept has been expressed historically through the disciplines of economics, sociology, politics and environmental social sciences. This approach provides a useful theoretical synthesis, but it also highlights the complexity surrounding the concept. Silbergh presents thirty propositions on sustainable development. Although the achievement of sustainability is compared with the pursuit of the Holy Grail, decision-makers at all levels are not excused from making a practical contribution towards it. There may be uncertainty about the destination, but the direction should be clear. A theoretical treatment of sustainable development is not the stuff of popular politics. Once unpacked however, its constituent parts should make a great deal of sense to ordinary people. The challenge for Scottish policy-makers is to make the issues meaningful on a day-to-day basis.

Public and Political Opinion

In the second section, the authors provide a contextual map of public and political opinion. A progressive agenda for change will not reach very far if it conceived without a firm grasp of the appropriate mix of incentive and regulation. Neither green parties, politicians nor consumer movements have yet achieved a step change in how environmental sustainability is approached. Energy and commitment have not been matched by lasting achievements. More effective means of dialogue between policy-makers and the public are essential if the modest advances achieved so far are to be built upon.

Chapter Three explores the pattern of environmental beliefs and behaviour in Scotland. It reviews some of the earlier findings of British and Scottish survey research and draws on the results of a Scottish opinion survey commissioned for the *Environment Scotland* project (and published for the first time in this collection). The authors note the limitations of opinion polling as well as the advantages of clearly 'costed' questions asked on a consistent basis over time. The evidence from previous studies of Scotland's position relative to the rest of Britain is inconclusive. While some have found Scots to be a relatively paler shade of green, others have noted the particular concern about nuclear power and waste, while others still have noted that Scottish consumers are more likely than the average to engage in some forms of green behaviour.

More revealing is the pattern of attitudes and behaviour within Scotland, derived from the 1998 survey. For example, there is strong support for developing alternative energy sources even if it costs more, while opinions are evenly split on the value of extra jobs at the expense of the greenbelt. Across a series of questions, educational attainment is the most powerful variable in distinguishing between both attitudes and behaviour. The higher the level of education, the more common are green beliefs and actions, and the gap is greatest for those options which are generally less popular. Less significant differences, although still notable, emerge between socio-economic groups (skilled non-manual workers are 'more green' than those above and below them on the occupational ladder); regions (the East is ahead of the West); and among 'car rich' households. As important are the associations which do not emerge. Youngest respondents are less likely than the oldest to take part in those environmental activities specified, while there is no gender gap to speak of in terms of green behaviour.

The chapter concludes with a call to Scotland's new policy-makers to 'make the sustainable choices easier choices - in other words, more accessible and affordable'. As part of this approach, the costs of reform must be clearly located and the benefits must be apparent if public support is to be secured.

'If support for the Green Party was used to measure the strength of environmentalism, one would be forced to conclude that green politics in Scotland is essentially irrelevant.' Lynn Bennie argues in Chapter Four that such a conclusion would be mistaken. She traces why Scotland's electoral climate has proved so cool for the Green Party although not for green politics in the broader sense. The Scottish party's lack of success, even compared with the fortunes of the Greens in England, is striking. It has been 'almost completely marginalised in recent political history. This is also contrasted with the domination of green politics by environmental pressure groups. While the Greens have played their part in local campaigning, political parties also exist to win influence through elections. In trying to be a pressure group and a party, Scottish Greens have found themselves uneasily located between the two: unable to convince voters to overcome their party attachments, but 'reluctant to associate with the more radical activities of the environment movement for fear of discouraging potential voters.' Bennie suggests this is a no-win situation, with the party neither credible enough to elect representatives nor sufficiently 'media-genic' to attract publicity.

Why has there been no breakthrough for the Green Party? The first-past-the-post voting system is without doubt a barrier to small parties whose support is low and spatially dispersed. Ireland's Green Party has won parliamentary seats on less than 3 per cent of the popular vote due to its system of proportional representation (PR). However, the Greens in England have managed to elect councillors under the same majoritarian system as in Scotland (although multi-member wards in much of England offers a modest advantage). A more convincing explanation is to be found in the degree of party competition in Scotland, where the presence of the SNP provides disaffected supporters of the other main parties with an alternative that does not exist south of the border. There is simply less political space available for the Scottish Greens to carve out a distinctive niche. The tentative formation of the independent Highland Alliance will make for even more of a squeeze.

The Greens look to PR to deliver a handful of seats in the first Parliament, but Bennie argues that electoral reform offers false hope for Scotland's Greens. Aside from the external barriers - including a regional threshold which will probably still be too high for the party to exceed - internal party factors must also be considered. 'The party's internal organisation was, for a period, nothing short of shambolic' and although it has been effective in working with other bodies on constitutional reform, it has yet to prove that it could work successfully with other *parties*. This makes sobering reading for those who wish to see a healthy green presence at Holyrood, but it also suggests that some of the necessary steps in the short-term can be taken by the party itself. Tactical resourcing and raising the profile of a small number of 'identifiable' candidates on a small number of salient issues may go against the grain, but is probably the only way of organising support in order to win any representation.

Policy Debates

The third section explores the substantive policy issues which are likely to occupy Scotland's new politicians and policy-thinkers in the early years of devolution. Is sustainable development more than old whisky in a new bottle? An immediate challenge will be posed by grassroots campaigns which tend to present politicians with difficult tradeoffs. Can the responsibility of the local authority or enterprise company to boost employment be reconciled with the interests of diverse communities? How should today's apparent advantages be judged against longer-term risks? And where local initiative goes beyond environmental protection to new forms of innovation, how can policy-makers provide the appropriate degree of support without exercising too much control?

In Chapter Five, Eleanor McDowell and Douglas Chalmers focus on the local arena to examine the nature of grassroots activity in urban Scotland. Three local initiatives are considered in order to explore the purpose and achievements behind such activity. Sustainable development is, ultimately, a challenge to the traditional model of governance (based on the assumption that policy elites based in central government have the capacity and authority to prescribe standardised solutions across all localities). It demands community participation and inter-agency partnerships in keeping with the overall philosophy of Local Agenda 21. The authors state: 'the greater the community involvement, the greater the erosion of the democratic deficit' in pursuit of development that is 'economically efficient, socially equitable and environmentally sound.' Associated concepts of quality of life and empowerment should guide thinking and action. Intermediate agencies standing at arms length from government and the private sector have a vital role to play in responding to local priorities quickly and in a more flexible manner, but local government has a particularly important contribution to make in fulfilling the objectives of LA21. Scrutiny of local authority commitment to sustainable development is likely to become sharper once the Parliament is elected. Councils will be expected to act upon their stated aim of decentralising environmental decision-making. They must also place a higher value on consensus-building across very different sectors if long-term change is to be secured.

The dilemmas around sustainable development have been played out most controversially on the territory of land-use planning and transport. The first of these will be shaped largely by negotiations between the Scottish Parliament and local government. The second is marked by the new boundary laid down by the Scotland Bill. While some powers over transport planning will be devolved, other strategic and financial powers will remain reserved to Westminster. Despite the powerful consensus in favour of an integrated transport policy and the generally favourable reaction to the Scottish White Paper of July 1998, the objectives may continue to prove elusive in the early years of devolution.

In separate essays, David Begg (Chapter Six) and Sarah Boyack (Chapter Seven) make the same point: local government reorganisation has reduced the capacity for coordination. According to Begg, the imposition of legislation encouraging a 'go-it-alone' mentality has had 'a devastating effect' on the ability of transport planners to address issues of strategy effectively. The new Parliament will inherit a country with more cars and more miles travelled by road each year. Yet the structure planning process which should allow for greater transport coordination across local authority boundaries is essentially a voluntary mechanism. Reorganisation in urban Scotland has carved out councils which are geographically too small to implement workable transport plans on their own.

In response, Begg notes that a quasi-regional tier of decision-making is necessary and feasible. Writing in the run-up to the Government's Integrated Transport White Paper, he favours five 'Transport Partnership areas' covering the city-regions and rural areas and notes that they should be given powers to pay for their respective transport priorities using different fiscal instruments, ranging from urban road pricing (as subsequently proposed in the Scottish White Paper), non-residential car parking taxes, and a share of revenues from the petrol price escalator. Partnership areas could provide an effective forum for joint planning by local councillors, Members of the Scottish Parliament and the appropriate officers. Begg concludes that such a structure is essential to reverse the fragmentation of recent years.

One specific concern, also alluded to by Dunion in the book's concluding chapter, relates to the Parliament's tax-variation powers. With discretion limited to the basic rate of income tax, the Government appears to treat taxation as simply a revenue-raising tool rather than one means of changing incentives and behaviour. He makes a convincing case for the Parliament to have the ability to vary taxes to reflect the different social and economic geography north of the border (such as raising petrol prices more slowly in rural Scotland). While there have been recent moves to permit local authorities to levy new transport charges, the author judges the Devolution White Paper to be 'behind the times in thinking on economic and environmental issues'.

Begg is, however, optimistic about the opportunities provided by devolution to make fundamental changes to the pattern of transport funding. To make a reality of an integrated transport strategy, the model of budgeting must change. An *integrated transport fund* is proposed, pooling existing sources of public and private funding, and the revenues accruing from new transport charges shared between road users, employers and consumers. Begg proposes a significant break with political tradition by making a strong case in favour of *hypothecation* of new funds to ensure the money is invested in the manner originally intended used for the purpose intended, and by recognising the need to phase in some of the benefits of a sustainable transport policy in advance of higher costs being levied.

The post-war planning system has for decades offered a window onto the policy priorities and ideology of governments. Sarah Boyack highlights the resilience of planning, having survived the harsh climate of the Thatcher years, only to be applied in the later Conservative years in defence of homeowners' interests and, to some extent, to promote environmental protection. Experience of the planning system is usually associated with disputes over proposed development of land or changes in its use. Such conflicts attract media attention precisely because they are untypical: fewer than 2 per cent of planning applications result in appeals. In most cases, for most of the time, planning is uncontroversial. The agenda for reform is concerned instead with the speed and effectiveness of the system in dealing with applications. The lag time between drawing up plans and their adoption typically exceeds five years. By the time they are implemented, they have too often fallen out of synch with the overarching structure plan. There is also growing interest in developing a series of planning services which anticipate change rather than getting stuck in fire-fighting.

The Scottish Parliament will add legislative competence to a planning system which has enjoyed substantial administrative devolution without actually exercising much of the discretion that has been available. Devolution will result in time for more effective scrutiny and for more legislation. But like Begg, Boyack notes that the capacity of the planning system is uneven and has been fragmented: 'Transparent, accountable and long-term decision-making is likely to become progressively more difficult to achieve as the impact of local government reorganisation works its way through the system.' While a *de facto* regional tier of land-use planning has emerged or survived in parts of Scotland, there are overall fewer staff to prepare, implement and monitor planning, at the same time as an increased need for liaison between the same sectors in different localities. Boyack identifies the need for a tier of decision-making between Holyrood and local government as one of the early issues for the Parliament to take a view on.

A more rigorous approach to monitoring trends might improve the system's ability to anticipate change rather than respond after the event. The author proposes an index composed of *State of the Environment* Indicators. While this would provide a national focus on sustainable development, it should also be able to accommodate the different planning priorities *within* rural Scotland as well as between remote, suburban and central city locations.

Boyack identifies two drivers for change in planning, in addition to the regional question. One is the development of fiscal instruments to shape where we live and how we travel to work, including levies on greenfield development and financial incentives for brownfield development (for affordable housing and to locate new sources of employment closer to where people do or could live). The second is the involvement of ordinary members of the public. Boyack expects devolution to give rise to a higher profile politics of planning, where urban/rural disputes may become more common. More effective conflict resolution processes should be developed in planning inquiries to encourage a problem-solving approach 'rather than stand-offs between opposing interests'.

One striking feature of the 1997 devolution referendum was the much higher levels of voting in favour in rural authorities than twenty years earlier. The Parliament is expected to take early action to strengthen the position of Scotland's rural communities. Andrew Raven explores the significance of agriculture and forestry in Chapter Eight, offering an explanation of the key influences shaping rural land-use patterns now and in the future. The chapter presents an *Audit* of the current situation, a *Vision* of what might change in the future, and a plan of *Action* to get there.

In mapping contemporary trends, Raven highlights the importance of state involvement in the Scottish rural economy. Government is the key player through the Common Agricultural Policy (which provides three pounds out of every four in agricultural subsidies) and forestry grants for example. Scottish farming is a business worth £1.8 billion; it employs almost 70,000 people directly, becoming more capital intensive while continuing to shed labour; Scottish farms are nearly ten times as large as the EU average; and four out of five pounds of financial support goes to just one in five farmers, helping to concentrate ownership of more agricultural land in fewer hands. Farming has also become a more complex and bureaucratic business: 'every field now has a unique map code number, every cow its own passport' due to EU requirements. One counter-trend is crofting which remains labour intensive and maintains a living and working environment in some of Europe's least productive agricultural lands. Despite rising productivity, not least in forestry where output has more than doubled in the last twenty years, all is not well in rural Scotland. Raven states that 'biodiversity has continued its remorseless decline'. Conservation has relied for too long on weak instruments on a voluntary basis. Incentives for sustainable land management must be more widely available and actively promoted.

An integral part of the sustainable vision is to overcome the sectoralism in rural affairs. Specialisation has been bought at the expense of capacity to see the big picture. The inter-departmental Committee for Rural Affairs creates a starting point for something closer to holistic government in rural Scotland. Its deliberations and to a greater extent those of the Parliament should be guided by a search for the greatest public benefit in the long-term. The plan of action should include a long-term reshaping of public subsidies: support for agriculture should 'mainstream' environmental priorities rather than add them on as an afterthought. Moreover, if the target of improving biodiversity is taken seriously, Raven argues that more public spending will be required. A brief but thorough code of good land use practice is essential. Without it, it will continue to prove difficult to measure improvements in the quality of land management, or draw conclusions on the balance between public and private returns. Raven offers a realistic account of the barriers and interests to be addressed. His conclusion is shared by a number of contributors: the prospects for reform are bright if the Parliament demonstrates its commitment (and ability) to act.

It is followed by a related essay on land reform by Hugh Raven (Chapter Nine). The chapter presents a clear account of the historical tensions surrounding ownership and maps out a series of options for reform. Land reform will be an early priority for the new Parliament. The body of law governing ownership and land-based employment has hardly changed through the centuries, making land reform as significant an issue in Scotland as anywhere in Europe. Raven notes that although the bulk of finance available for public interest land management will not be within the Edinburgh's control, *land tenure* will be a wholly devolved issue and provides the obvious starting point for a programme of land reform.

Hugh Raven underlines a theme running through a number of the contributions in the book: the need for more information, of a better standard, more widely accessible to the public. Two areas are identified as priorities. First, information about the pattern of land ownership is very uneven, depending on a combination of local and historical factors. A register of all land holdings above a specified size should therefore be published 'as a minimum requirement'. Second, Raven calls for an end to the obsession with secrecy around the distribution of government subsidies for land. Those who receive public money to manage land should be required to justify how it is used. There is no case for withholding such information on the spurious grounds of commercial confidentiality.

As part of a strategic overhaul of land management, a Land Commission could be established with powers to tackle poor management and prevent land falling into mis-management in future. Clearer conditions of eligibility to purchase land would be needed. While 'carrots' for good management can raise standards, 'sticks' must also be available and used when appropriate: 'There is widespread dismay at the reluctance of public authorities to use compulsory purchase powers even where there is clear evidence of abuse'. Compulsory purchase orders, backed up by a range of ownership options, are essential components of Scottish land reform. A power is not a power without the commitment to use it wisely. Greater opportunities for tenants to take crofting estates into community ownership at low cost may prove particularly attractive in remote areas, although community ownership in itself is no guarantee of success. Long-term community interest, organisation and capacity for management must be demonstrated rather than simply assumed. The chapter concludes with a practical eight-point plan for land reform which the Parliament could begin work on immediately.

Energy policies have been closely tied to debates about economic development, fuel poverty and Scotland's position within the United Kingdom itself (at least since the early 1970s campaign declaring 'Its Scotland's Oil'). While disputes over the future ownership and the potential of one energy sector may continue, a sustainable energy policy must range more widely. Tony Gloyne and Alan Hutton trace the shift in government instruments from ownership (nationalised energy corporations) to competition, regulation and pricing policies (Chapter Ten).

After more than two decades of taking about the potential for alternative energy in Scotland, the authors argue that the time is right for cross-sectoral planning to be given a higher priority. Scotland is an energy-rich country. Like other developed economies, it relies on large inputs of non-renewable energy sources (fossil and fissile fuels). Only around 10 per cent of electricity is generated from renewables. Such a pattern of energy production is 'inherently unsustainable'. Rising levels of carbon dioxide are predicted. An alternative measure of Scotland's economic welfare suggests that more wealth, created in the ways it currently is, will not buy a higher average quality of life. The authors argue that the transition to an energy base with a very different composition needs to be made in a relatively short time period. There will be two clear strands to a more sustainable energy policy: a larger stock of renewable energy; and greater efforts to develop the 'fifth fuel' of energy conservation.

There is no doubt that Scotland's renewable energy potential is high, both on-shore and off-shore. But the economic potential of renewable energy is less clear. Wind power could supply about half of Scotland's current electricity supply, but at uncertain cost. The authors favour the establishment of a Renewable Energy Development Agency to help stimulate growth in the relatively new environmental technologies sector. In common with other authors in this book, Gloyne and Hutton believe that a 'fundamental shift in the base of taxation away from income and towards energy and resource use is desirable,' although they are sceptical about the ability of the Scottish Parliament to make tax changes which are not shared by the rest of Britain. Understandable caution in the early years of devolution need not prevent revenue-neutral changes in the composition of the tax base being introduced, starting perhaps with a non-domestic energy tax partly replacing the fixed property tax facing businesses. Gloyne and Hutton do not make this suggestion in their chapter, although we believe it follows from their analysis.

Greater energy conservation will offer Scotland significant returns. Improvements in the energy efficiency of the housing stock should be a high priority. The Parliament could set a target for Scottish housing to be cheaper to heat and warmer (approaching standards achieved in Sweden despite its harsher winter climate) by the end of its first term.

The focus of regulation in the energy utilities has been on reducing end-user energy prices as the main measure of benefit. As market liberalisation proceeds, the case for better regulation becomes more compelling not less so. The authors argue that consumer energy costs should be reduced by energy efficiency (and thus less need to burn fuel) than through the artificial reduction of unit prices. Gloyne and Hutton argue that: 'This represents an impediment to the development of a sustainable energy strategy for Scotland. There appears to be nothing in principle that should prevent the devolution of regulatory responsibilities....We take the view that the presumptions of current regulatory practice are inadequate or inappropriate'. Even if powers to regulate energy strategy are to remain at Westminster, energy-related policies will be within the remit of the Scottish Parliament. As with transport policy, the boundary between devolved and reserved powers around energy may in any event turn out to be moveable rather than carved in stone.

Sustainability is too complex to be secured by pulling the right policy levers at Holyrood, even if the level of coordination in government is markedly improved under devolution. New models of governance must share responsibility for sustainable development with the second and third sectors of business and voluntary/community organisations. This section of the book therefore concludes with two perspectives on the contribution of businesses to sustainable development.

Alistair Dalziel presents a concise and convincing argument in Chapter Eleven based on his business experience of environmental audit. While compliance with legislation is an integral part of practising greener business, a system of beliefs and values is also required. Green thinking and behaviour in the company cannot be tagged on as an afterthought. Practical reasons for companies pursuing a greener image include reputation and customer loyalty. Examples of good practice range from the minimisation of waste to energy conservation and environmental auditing. The inability to address such initiatives may result in financial penalties being imposed by legislation or competitive pressures being imposed by more successful and enlightened companies (whose customers may 'select against' those which fail to comply). Dalziel concludes that adopting greener business methods can enhance a company's effectiveness: 'good environmental management means good business.'

Continuing the theme of business environment strategies in Chapter Twelve, Peter Strachan takes the view that 'command and control' legislation has traditionally been considered one of the most effective ways to improve environment performance standards in companies. This approach has been succeeded by a more managerial model. Environmental Management Systems (EMS) are gaining increasing interest in government and industry-led bodies. The author suggests that participants in EMS have to 'fundamentally rethink their core business activities in order to make them more consistent with the macro goals of sustainable development.' A qualitative analysis based on a survey of all British organisations participating in EMS identifies some implementation problems (or 'hotspots'). Strachan concludes with a series of practical recommendations for the Scottish Parliament to consider, with a view to raising awareness and promoting wider implementation of EMS.

Scotland, Europe and Beyond

The fourth section presents the book's concluding chapter. Rather than providing an overall conclusion drawing together the various propositions and recommendations of the authors, each contribution stands in its own right. The authors were not charged with the responsibility of coming to an agreement on key conclusions. Chapter Thirteen instead goes beyond the questions threaded through the rest of the book to consider Scotland's role on the international stage. How does a small country within a union-state and the European Union fulfil its wider responsibilities? Could it contribute more than it is bound to as a constituent member of the UK or EU? How will it adjust to the results of UK negotiations on the world stage? And will Scotland be a pioneer in raising the standards of environmental protection, or a laggard, acting only after it has been forced to by Brussels?

In a far-sighted analysis, Dunion explores the realpolitik of a new Parliament adjusting to the constraints in which it will operate on a UK, European and international stage. His essay poses a series of provocative questions, in contrast to various other accounts of devolved policy-making which are either over-generalised or over-optimistic. His sober assessment of the risks and opportunities of devolution notes some of the political and financial strains which may emerge: 'The Scottish Parliament will inherit responsibility at the very moment when states, far less sub-states, have a markedly diminished ability to strike a distinctive set of policies'. The Parliament will be subject to forces from within Europe and particularly from further afield with limited room for influencing the framework. Dunion believes that powerful unaccountable forces ranged against the powers of a newly-elected legislature will give rise to frustration. While opportunities for influence will be limited, 'there is absolutely no reason why the Parliament should not be represented on UK Government delegations to the United Nations for example.'

Echoing the concern of a number of contributors to the book, Dunion believes the tightly-prescribed fiscal powers of the Parliament will quickly come to be seen as problematic: 'This may turn out to be an ill thought-out and unnecessarily narrow financial settlement' He also identifies tensions over the environmental responsibilities inherited by the Parliament if future adjustments to Scotland's share of UK public spending fail to take account of the diversity and scale of the country's responsibilities.

The hotly disputed Multilateral Agreement on Investments (MAI) may seem to have been negotiated far from the circles of Scottish influence, but its impact will nevertheless be widespread. Although the OECD

(Organisation for Economic Cooperation and Development) denies the charge that MAI will frustrate prospects for sustainability, the author notes the strong perception 'that environmental and labour standards would always be optional extras to be disposed of in favour of maintaining free trade and globalisation', citing the non-binding nature of environmental and social protection clauses compared with stronger mechanisms upholding companies' economic rights. Dunion argues it would be almost impossible under the MAI for Scotland to introduce tougher criteria on land ownership, since new restrictions might 'unfairly' prejudice a foreign investor compared to a Scottish investor. Unlike the example of Denmark, Scotland cannot benefit from pre-emptive action to safeguard its forthcoming land legislation.

Unlike most discussions at the UN or the OECD, the direct consequences of decisions taken within the EU will be felt in Scotland, often at a very local level. Although there are reasons to be hopeful that Scotland's new politicians will wish to fulfil their responsibilities on environmental protection, sustainable development is not always a badge to be worn with ease. The Scottish Parliament will inherit 'a legacy of under-investment in environmental protection' which may cause it to fall foul of European Directives. There are risks that measures to do more to protect the environment will be perceived as having an adverse effect on jobs, business costs and investment decisions. There is concern that Scotland will drag its feet on implementing EU Directives (on bathing waters for example) and obstructing tighter laws in the future. If 'a reputation for simply whingeing or being environmentally reactionary' is to be avoided, Scottish interests must be represented at an early stage in formulating the UK position. The task will be to mitigate negative impacts without denying the principles of sustainability.

If the Parliament does act in a parochial manner, it can expect to be charged if it fails to meet its obligations. But equally it should bear no part of any fine associated with Westminster's failure to act. Concluding on a cautiously optimistic note, Dunion argues that Scotland has an opportunity to move further and faster than Westminster: 'Acting to secure higher air quality, better bathing water, more stringent planning or building regulations would of course mean adopting different standards north and south of the border. The capacity to do so is the whole point of devolution'.

2 Sustainability and Sustainable Development

DAVID SILBERGH

> The principal defect of the industrial way of life is that it is not sustainable. (Goldsmith *et al*, 1972, p.2)

In simple terms, sustainable development is usually regarded as a marriage of environmental and economic concerns, a form of economic development which does not cause excessive environmental damage. 'Sustainability' on the other hand, is a less well defined concept, which tends to be prefixed with 'economic', 'social' or 'environmental' for clarity. Both terms are much used today in academic and policy circles and are slowly becoming part of everyday parlance. However, they are terms which are often misused. A simple analogy of the terms would be to liken the concept of sustainability to the Holy Grail and to liken the drive for sustainable development as the search for it. This chapter therefore seeks to explore some of the theoretical background to sustainability and sustainable development which in turn underpins the public policy framework within which organisations and individuals undertake actions in pursuit of sustainability and sustainable development.

How are sustainability and sustainable development defined? The Shorter Oxford English Dictionary does not devote an entry to either, but gives a hint of that which is sustainable being 'supportable; maintainable'. This is a basic definition, but due to its simplicity is probably as good as any other. There are a huge number of definitions of sustainability, and a selection of these are reproduced as Annex One. However, for the purpose of this chapter, sustainability will be defined in accordance with *Caring for the Earth* (IUCN/UNEP/WWF, 1991). That is, sustainability can be considered, 'A characteristic of a process or state that can be maintained indefinitely.'

It proves very difficult to properly analyse public policies for sustainability. Although the key documents commonly make reference to the term, they regularly fail to define it (q.v. Annex One). Thus, the focus of the research referred to was on sustainable development which is 'development that can be maintained/supported for the foreseeable future (in economic, social and environmental terms)' (Oxford Dictionary). Even if the Holy Grail's existence is questionable, that in itself need not prevent us from searching for it. It is also debatable whether the American or British systems of democracy can truly deliver Abraham Lincoln's dream of 'government of the people, by the people, for the people,' but this would not generally be regarded as a reason for turning our backs on the democratic ideal.

Sustainable development is usually expressed in a short-hand way as being that which:

> ....meets the needs of the present without compromising the ability of future generations to meet their own needs (Report of the Bruntland Commission, *Our Common Future*, 1987, p.8).

The story of sustainable development did not begin in 1987. The term was used as early as 1972 in *Blueprint for Survival*, a special issue of *The Ecologist* magazine. Since then it has become more widely used, and defined in diverse ways. Indeed a great deal of the material which is written about sustainable development serves only to muddy the conceptual waters further. In this chapter, the Brundtland Commission's definition of sustainable development is adopted. Whilst there may be arguments for and against its use, it has been selected because:

(a) regardless of whether one agrees with it or not, people throughout the world are familiar with it. It serves as a benchmark against which the vast majority of actions or writings completed in the name of sustainable development can be analysed;
(b) in terms of assessing aspects of governmental policy in the UK it is useful because the Brundtland definition was adopted by the Government as a basis for the *UK Strategy for Sustainable Development* (Secretary of State for the Environment *et al*, 1994a)[1].

To further clarify the Brundtland definition, sustainable development has been defined by Kumar (1993) as a complex concept comprised of concern for the environment, the economy and society. Thus, sustainable development should be regarded as a drive towards meeting, 'the needs of the present without compromising the ability of future generations to meet their own needs,' be these needs social, economic or environmental. Put another way, Rubenstein (1994) refers to the 'three Es' of sustainable development. He defines it as 'a moving, dynamic balance between economic sustainability, equity of distribution, and ecological sustainability.' Searching the wealth of writing on the environment, the economy and society was therefore a necessary precursor to defining sustainable development. This approach was not idiosyncratic however, with the Secretary of State for the Environment *et al* (1994a) declaring that to begin to comprehend the nature of sustainable development, one must first appreciate that, 'Understanding how the economy, society and the environment work and interact is vital in predicting the nature of change, its consequences and how this may be turned to our advantage.' Furthermore, it is no exaggeration to state that the literature on the subject is probably now being produced at a rate greater than that at which it can even be catalogued, let alone reviewed. In seeking to define sustainable development therefore, a review of the socio-economic, environmental and 'sustainability and sustainable development' literature was undertaken. Given the confines of space however, a decision was made to focus this chapter on identifying those elements of the complex concept of sustainable development which can be traced to origins in the socio-economic literature.

Economy and Society

In economic and social terms, traces of what sustainable development is about can be seen through a great variety of works. For example, in the last century, Marx recognised the major resources which underpin society as we know it in modern times as being land, labour and capital. Writing with Engels (1965) in *The German Ideology,* Marx argued:

> In communist society, where nobody has one exclusive sphere of activity but each can become accomplished in any branch he wishes, society regulates the general production and thus makes it possible for me to do one thing today and another tomorrow, to hunt in the morning, fish in the afternoon, rear cattle in the evening, criticise after dinner, just as I have a mind, without ever becoming hunter, fisherman, cowherd or critic (pp.44-45)

While Marx viewed the greatest threat to human society as the mis-use of capital and the over-exploitation of labour, his ideas would not be too difficult to reconcile with modern notions of sustainable development had he paid as much attention to the over-exploitation of the environment as to the structure of land ownership. That is, sustainable development is a process very much concerned with:

- rejecting large scale industrial production (i.e. shifting the economy from one dominated by large-scale industry, be it run by individual capitalists as in Marx's day or by multi-national companies as it is today);
- bringing an end to the over-exploitation of people, although those suffering most hardship today live in countries Marx had probably never heard of;
- protecting the environment (which at a simple level consists of land, sea and atmosphere).

A further (non-Marxist) social vision with implications for sustainable development was expounded in Ebenezer Howard's book *Garden Cities of To-Morrow*. Originally published as *To-morrow: A Peaceful Path to Real Reform* in 1898 and in a revised edition under the current title in 1902, it is widely regarded as a precursor to Frederic Osborn's famous *Green Belt Cities*, so influential in shaping our environment in the Post-War era. Howard was deeply concerned with the depopulation of the countryside, the attendant decline in agriculture and with the unsustainable growth of urban settlements which were lacking in green space, becoming ever more polluted and populated with people living in poverty and squalor. Even at the turn of the century, Howard was acutely aware of the rôle which technology had to play in shaping human welfare. His perception of technology as a tool which can be put both to useful and damaging ends on a global scale is a perception of some significance even (or especially) in today's world. Howard wrote (1965 edition):

> What a happy day it would be for the people of this country, and of all countries, if they could learn, from practical experience, that machinery can be used on an extended scale to *give* employment as well as *to take it away* - to *implace* labour as well as to *displace* it - to free men as well as to *enslave* them (pp.79-80).

Therefore, another clue to the meaning of sustainable development is that the application of scientific and social scientific knowledge through appropriate technologies will be essential for the well-being of our planetary environment, economy and society. Conversely, great harm can be caused by technology developed or used in an inappropriate fashion, for example nuclear fusion will theoretically give mankind an immense source of clean and safe energy. So far its use has only been perfected in the manufacture of the hydrogen bomb, the greatest weapon of mass destruction ever developed. Sustainable development is about finding the balance, not about technology *per se*, but about making value judgements on its use (see also: Clarke and Roome, 1993; Gault, 1993; Kemp, 1993; Skirbekk (ed.), 1994).

Thorstein Veblen, writing in *The Theory of the Leisure Class: An Economic Study of Institutions* (1899), was a strong critic of the harm to society which he saw as being caused by inappropriate use of technology. In the same way that Howard's inspiration of Osborn came to heavily influence the shape of our planning system, Veblen's work was to influence what is one of the most important social scientific analyses of tourism this Century, Professor Dean MacCannell's *The Tourist: A New Theory of the Leisure Class* (1976). Veblen, like Marx, believed in a bygone Golden Age, a time when man was unfettered by the material (similar to Marx's concept of 'primitive communism'). He wrote that in this early state of development:

> Leisure...came to hold a rank very much above wasteful consumption of goods, both as a direct exponent of wealth and as an element in the standard of decency, during the quasi-peacable culture. From that point onward, consumption has gained ground, until, at present, it unquestionably holds the primacy...(pp.91-92).

Veblen uses the notion of, 'the canon of conspicuous waste ' to explain the consumption patterns and thereby the problems of modern society. He argued that people in modern society accrue status from blatantly wasteful expenditure, be that the employment of servants or the replacement of a refrigerator because its colour no longer fits with the new wallpaper. Under such norms, where profligacy is viewed as desirable, man's ability to pursue it is abundant, to the extent that, 'The need of conspicuous waste, therefore, stands ready to absorb any increase in the community's industrial efficiency or output of goods...'

Furthermore, whilst the criticism of 'conspicuous waste' is broadly compatible with Marxist criticism of the needlessness of opulence enjoyed by the capitalist class, there are hints in Veblen's book that he had gone beyond the rejection of conspicuous waste on grounds of social and economic equity. It is apparent that Veblen understood the dangers which over-consumption held for the environment, pressure on which could ultimately lead to social and economic collapse (see Tainter, 1988; Clayton and Radcliffe, 1996). Veblen notes that:

> The forces which have shaped the development of human life and of social structure are no doubt ultimately reducible to terms of living tissue and material environment...

and continues:

> It is only as the exigencies of modern industrial life have enforced the recognition of causal sequence in the practical contact of mankind with their environment, that men have come to systematise the phenomena of this environment, and the facts of their own contract with it, in terms of causal sequence.

Veblen's study, published in 1899, provides us with further clues to unravelling the secrets of developing sustainably. Such development:

- is inextricably linked to the wise use of resources in the pursuit of economic well-being. The consumption of resources in a profligate manner will add little to our development as a society or to our individual quality of life;
- recognises humanity's ultimate dependence on the health of all aspects of our environment, not only the land, sea and atmosphere (living tissue) as previously described, but also wholly material and modified environments.

Sustainable development is therefore about maintaining the vitality of our environment in order that we may maintain the vitality of our civilisation. However, it is a matter of great concern that eighty-nine years after Veblen wrote: 'The need of conspicuous waste, therefore, stands ready to absorb any increase in the community's industrial efficiency or output of goods..', the world was one in which conspicuous waste was alive and kicking, evident in the following quote from Tainter (see also Durning, 1992; Redclift, 1996):

> ...the productivity of industrialism for producing social welfare is declining...while US per capita product increased 75 per cent from 1950 to 1977, weekly work hours declined by only 9.5 per cent (Tainter, 1988).

Perhaps this is not so surprising. After all, John Kenneth Galbraith reiterates in *The Affluent Society* (1958) Veblen's belief that, ' ...both the moral and material debasement of man, was part of the system and would become worse with progress.' Galbraith adds that not only are we becoming worse at using the resources at our disposal, moving from:

> ... a world where more production meant more food for the hungry, more clothing for the cold, and more homes for the homeless to a world where increased output satisfies the craving for more elegant cars, more exotic food, more erotic clothing, more elaborate entertainment - indeed for the entire modern range of sensuous, edifying, and lethal desires (p.109).

but that we are failing to rectify the mis-use of resources in the marketplace through the extensive state apparatus which we have established to do precisely that: 'We exploit but poorly the opportunity along the dimension of public services. The economy is geared to the least urgent set of human wants.' Galbraith blames this predicament on what he calls 'the paramount position of production' and devotes most of his 1958 publication railing against it, particularly in modern Western civilisation, where he felt that the predicament is particularly pernicious and evident on a truly monumental scale:

> The poor man has always a precise view of his problem and its remedy: he hasn't enough and he needs more. The rich man can assume or imagine a much greater variety of ills and he will be correspondingly less certain of their remedy. Also, until he learns to live with his wealth, he will have a well-observed tendency to put it to the wrong purposes or otherwise to make himself foolish. As with individuals so with nations (p.1).

Whilst Galbraith has modified his views during the forty years which have passed since the publication of *The Affluent Society*, he is still firmly of the opinion that we fail to use wisely the resources at our disposal and that we are failing to rectify this mis-use of resources through the apparatus of the State (Galbraith, 1996).

The concept of sustainable development has already been alluded to in terms of its planetary dimension. The above quote from Galbraith is again hinting at the international dimension of the economy which is in fact a key consideration in any discussion of sustainable development. That is, the concept of sustainable development addresses issues on a macro scale. There is little point in aiming to develop sustainably in only one town or country if the rest of the world community continues to consume and pollute with gay abandon. This is not to say that all towns and countries can invoke the same tactics to develop in a sustainable fashion (although they should undoubtedly act in pursuit of the same macro goal, i.e. global sustainable development and the rejection of reckless waste). Rather, as Galbraith suggests, rich and poor nations face problems of varying degrees of severity and immediacy, with the poor countries needing a larger slice of the global economic cake and the rich ones needing to learn to live with their wealth and put it to better use. Thus, sustainable development is a call to international improvement in the health and wealth of people and their environment, with a special focus on the needs of the poor nations. While this chapter deals neither with the issues faced by impoverished nations nor their relationship with the industrialised world, it is meaningless to speak of sustainable development without at least cursory reference to them.

The need for the developed world to encourage and assist poorer nations in their search for sustainable development becomes ever more necessary. The speed at which economies and societies outwith Europe, North America and Australasia are developing from a primarily agrarian (or even hunter-gather) base towards industrialisation continues to increase exponentially, as described by Rostow in his seminal 1960 text *The Stages of Economic Growth: A Non-Communist Manifesto*. The central thesis of Rostow's book was ground-breaking, (written in the middle of the Cold War). He proposed that economic development would follow a common pattern *regardless* of whether an economic system was operated under a communist or capitalist political system. Hence the title of the book. Whilst Rostow agreed with Marx that all economic systems will follow a common pattern of development, he renounced the theory that this was inextricably linked to a necessary pattern of social and political development: he refuted both the Marxist belief that economies could not fully mature without a socialist revolution and the conventional wisdom of capitalism which contends that only a privately owned and relatively unfettered economy can provide the means to achieve a civilised and industrialised nation. The gravest danger which Rostow saw for the development of the global economy was not related to polity, but to the speed at which the economies of poorer nations were developing. The United Kingdom took hundreds of years to move from a feudal system to today's high consumption society, while some poor nations are now moving from even pre-feudal systems to high consumption systems in a matter of decades.

Such economic transformations tend to lead not only to massive political and social unrest, but to a huge degree of environmental degradation such as deforestation on a massive scale, increasing toxic discharges to atmospheric and aquatic environments, and the rapid growth in slums. Since Rostow's book was published the pace of development in many poor countries has continued to accelerate, along with the attendant environmental degradation.

Finally, with respect to the developing world, it is essential to note that global sustainable development will result not only from instruments of foreign policy, but with the implementation of many domestic policies. Initiatives at home, whether undertaken by Government, business or individuals should be completed with regard to the effects these actions will have beyond our national boundary, and especially in the Third World.

Thus far, the social and economic theory from which the picture of sustainable development is being painted has only been concerned with the environment in an implicit sense. The works of Marx were inspired by his abhorrence of the alienation and exploitation of industrial workers, the endeavours of Howard were motivated by the growth of slums and those of Veblen and Galbraith by a fundamental objection to the patterns of production, consumption and waste in modern Western society. That is not to say that these works do not contain several of the key elements of sustainable development, just that they were not developed with much regard to explicit environmental concerns. However, from the late 1960s onwards, a new type of analysis was beginning to emerge in the social sciences, concerned with environmental affairs as the primary object of investigation. Primarily it was in the field of economics rather than politics or sociology that this interest in matters environmental began, although these other disciplines were later to follow the lead of economists such as Kenneth Boulding and Edward Mishan.

In the Foreword to his 1967 book *The Costs of Economic Growth*, Mishan notes that '...bringing the Jerusalem of economic growth to England's green and pleasant land has so far conspicuously reduced both the greenness and the pleasantness...' and suggests that, '...doubts about a positive connection between social welfare and the index of economic growth are amply justified.' The central tenet of Mishan's thesis is that Western society has become almost addicted to continued economic growth, and that there is a valid basis in neither economic theory nor experience for this situation to persist. For example, in the writings of Adam Smith (*The Wealth of Nations*, 1982, originally published 1776), it is quite plain that he did not expect economies to grow and grow. Rather, through his observations of agricultural output and its relationship to the price of grain, Smith expected economic systems to be in an almost perpetual state of flux, a state in which external factors impacting on the laws of supply and demand would lead to periods of both boom and bust. Thus, Mishan's contention is not only that the continued economic growth scenario is untenable, but that for the first time in mankind's history it has come to be expected. Mishan argues that the period of rapid economic recovery followed by annual growth which occurred after the Second World War came to be seen as a start on the road to never-ending economic growth, rather than as a period of massive restructuring and modernisation in the economic sphere. Prior to the last War, Mishan argues, nobody had even contemplated an economic system which could simply continue to accelerate. The 1920s and 1930s were decades in which grave economic decline affected Europe and North America. However, since the late 1940s, politicians, the public and economists have become so well acquainted with the notion of annual economic growth that they can no longer imagine any alternative. Of course, economic growth in an industrial world is powered by non-renewable fossil fuels and results in the consumption and waste of vast quantities of other non-renewable resources. Thus, advancing economic growth has become almost inextricable from the advancing degradation of the environment. Mishan expounds that:

> the adoption of economic growth as a primary aim of policy, whether it is urged upon us as a moral duty to the rest of the world or as a duty to posterity, or as a condition of survival, seems on reflection as likely to add at least as much 'ill-fare' as welfare to society (p.37).

and that the remedy for this is to be found in pursuing development in line with his stated belief that:

> the chief sources of social welfare are not to be found in economic growth *per se*, but in a far more selective form of development which must include a radical reshaping of our physical environment with the needs of pleasant living, and not the needs of traffic or industry, foremost in mind (p.8).

The concept of sustainable development can be said to wholeheartedly encompass Mishan's words. It was previously suggested in discussing Veblen that such development, 'is inextricably linked to the wise use of resources in our pursuit of economic well-being'. This statement can now be further developed in the light of Mishan's criticism of the cult of continued economic growth. That is, sustainable development, whilst not entirely incompatible with economic growth, cannot be achieved at the same time as year-on-year economic growth. Year-on-year economic growth in an industrial society organised as ours is today can, in the long run, only result in greater environmental degradation. It could be said that economic growth sustained over a lengthy period of time is nothing less than the antithesis of sustainable development where such growth is driven by the reckless consumption of non-renewable resources. Mishan's point was that it is illogical in economic terms to expect an economy to grow indefinitely. However, if we cannot displace this goal from the political agenda in the short to medium term, we should at least aim to encourage its demise in the long run and minimise the consumption and pollution of the natural resource base in the meantime.

Hodson (1972), writing in *The Diseconomies of Growth*, echoes the sentiment of Mishan in criticising the standard measurement of national wealth through the calculation of the Gross National Product and providing a perceptive note on the nature of growth:

> Growth is a biological term. Living things grow. Their growth may be good or bad: cancers are growths, so are dropsical swellings. A child may grow in height and weight while not growing in mind and understanding, or in mind and understanding while not in physical strength and proportions. The growth of a national economy, fully understood is like the growth of a living creature: it is made up of many elements, and it is not to be measured only by one set of money-expressed statistics (p.8).

Hodson continued to develop this theme of growth as a double-edged sword, in much the same way as Ebenezer Howard viewed technology. Warning of the irreversible damage which it can inflict upon the environment and the often unforeseen manner in which such damage occurs, he wrote: '...in the face of ignorance, caution is the rule. Drive slowly in fog.' Herein lies a clue to another essential trait of sustainable development, commonly referred to as *the precautionary principle*. Basically, as the damage caused to our planet and atmosphere by inappropriate development can sometimes be considered irreversible on anything other than geological timescales, sustainable development is about treading carefully. Where there is doubt over the likely extent of negative impacts resulting from an action, we should, if we wish to develop sustainably, avoid its execution until such time as we can more confidently predict its side-effects (see O'Riordan and Cameron eds, 1994).

Continuing on the theme of the dangers of unchecked economic growth, it is relevant to refer here to the work of Herman Daly, an economist whose work has significantly contributed to the emergence of the concept of sustainable development. In many writings on what he terms the 'steady state economy', Daly argues not only that such a 'steady state' is necessary, but proposes that it can be achieved with sufficient commitment to three fundamental principles: ensuring constant population levels; properly maintaining the planetary resource; and limiting the degrees of inequality in accessing the planetary resource which exist between differing human populations. Daly attempted to treat the relationship between the development of the economy and the state of the planet in a highly systematic manner, steering the economic discipline away from its traditional basic precept of the planet as a source of abundant raw materials which will be subjected to negative externalities (i.e. costs not accounted for through the pricing mechanism) as a result of processing these raw materials industrially.

Daly bolsters the criticism of waste as advanced by Veblen and Galbraith, adding a distinctly environmental objection to the essentially social objections of those who preceded. He also reiterates the argument put forward in discussion of Galbraith - sustainable development is about the ability of differing populations to adhere to the same principles and aims, to act in a multitude of ways to achieve these general aims, according to local circumstances. That is, to '...provide macro-stability while allowing for micro-variability, to combine the macro-static with the micro-dynamic,' (Daly, 1973). Twenty-five years later, Daly is still one of the leading authorities on environmental economics, although since 1973 his work has developed from the analysis of the effects of economic systems on the natural environment into a more wide-reaching theory of sustainable development, as can be seen in *For the Common Good...* (Daly and Cobb, 1990).

Perhaps the best known of the early socio-economic writings which addressed environmental concerns explicitly however is Fritz Schumacher's *Small is Beautiful: A Study of Economics as if People Mattered* published in 1974. Overall, Schumacher's work can be considered an economic text, but one which pays a good deal of attention to inter-linkages with social and environmental concerns. It is certainly less technical than many of the other well known environmental treatises of the time (such as *Limits to Growth*), but it is fair to say that although dated in some areas, it is still a thought-provoking book, the philosophical basis for which remains relevant. Schumacher founds his thesis on the concept of 'enough' stating, 'There are poor societies which have too little; but where is the rich society that says: 'Halt! We have enough'? There is none.' He then uses this concept of 'enough' to expound the virtue of organising our social and economic world in small units, an organisational model in which he appears to have passionately believed. This could be described as an exposition of the 'think global, act local' maxim. Schumacher called for his principle to be extended to nearly all aspects of human life, suggesting that: there are great benefits to be had from living in small rural communities; that there is much to be said for the operation of small business; the management of large organisations should wholeheartedly embrace the concept of subsidiarity in regard to decision-making; and the government's approach to economic development should address issues on a district or regional scale rather than a national one (he roundly rejected initiatives such as George Brown's *National Plan* of 1965). Thus, Schumacher could be said to hold the key to the *implementation* of sustainable development. Regardless of whether his 'smallness' criterion is referred to today as 'community empowerment' or as 'getting close to the customer', Schumacher's point is that the best way to develop our economy and society without detriment to the environment is to invoke the principle of subsidiarity and ensure that decision-making and action in all areas of our lives occur within the smallest possible unit that can effectively cope.

Dennis Pirages and Paul Ehrlich (1974) in their *Ark II: Social Response to Environmental Imperatives* also added to the stock of literature, focusing on the nature of the environmental problem as one which can only be solved by technological advance if it occurs in tandem with social change. They accentuate the need to foster what they term 'cultural evolution', proposing that:

> To direct cultural evolution is to make culture an effective weapon in the battle for human survival.....Modern industrial society can be described as technologically overdeveloped while remaining institutionally and socially underdeveloped. Science and technology have eliminated many of the problems of physical survival that have always confronted man, but as yet science has offered little advice germane to the settlement of the social problems that it has been instrumental in creating..... the United States is able to run the most complex of technological circuses but seems paralysed when called on to reach goals related to social change (p.44,pp.53-54).

However, Pirages and Ehrlich are vehement in their disdain for what they refer to as 'the new 'Bambiism' strongly criticising those aspects of 'New Age' thought which promote the idea of a return to a more primitive cultural state instead of pursuing a path of 'cultural evolution' temporally concurrent with technological advancement but in control of it. They state that:

> Many well-meaning people are eager to lead humanity to the happiness of the forest, forgetting that Bambi eventually saw the forest burn to the ground. The thought that any substantial portion of today's urban population could find contentment and sustenance communing with Nature in a shack in the woods is a romantic notion. Clearly, no religious answer for industrial society can be found in a retreat to the idyllic forest (p.281).

Pirages and Ehrlich stress that it would be as naïve to expect to move our world back to a prior state as it would be to bury our heads in the sand, voluntarily ostracise ourselves from society and pretend that all is well. We could well face serious social unrest if we were to attempt to recast our secular world of material comfort as a Victorian society, let alone return to the time before Eve ate the apple. Thus, any move towards sustainable development must be *gradual* but ongoing and based on achieving realistic targets, one at a time. It must also be accompanied by an honest, objective strategy of public education, which explains to the general public (who are more often than not very attached to their cars and domestic appliances) the reasons behind setting the targets to be achieved. A further precursor to developing in a sustainable manner is that any such public education strategy must be credible not to the faithful but to the unconverted. Dangers abound in this regard, as can be seen in George Orwell's damning indictment of socialists in *The Road to Wigan Pier* (1962). To what extent do these words hold true in today's context if one replaces the words 'socialist' and 'socialism' in the passage with 'environmentalist' and 'environmentalism'?

> I don't object to Socialism, but I do object to Socialists. Logically it is a poor argument, but it carries weight with many people. As with the Christian religion, the worst advertisement for Socialism is its adherents (p.152).

Whilst Orwell's analysis is far from academic, it holds what is perhaps the cornerstone of developing sustainably within it - we cannot develop sustainably simply through the issue of governmental edicts or business charters for the environment - we can only do so having captured the heart, soul and minds of ordinary people. We cannot expect honest, objective and reasoned arguments to be heeded unless advanced by persons *perceived* to be honest, objective and reasoned.

Hirsch's *Social Limits to Growth* (1977), like the work Pirages and Ehrlich, Daly, Hodson and Mishan, presents a valuable source of information on the social and economic characteristics which underpin environmental degradation. Hirsch refers to 'the paradox of affluence' as being the greatest threat to modern society and his book can probably best be regarded as a foil to the *Limits to Growth* report to the Club of Rome (Meadows *et al*, 1972). Containing a much more rigorous economic analysis than the 1972 Club of Rome document, Hirsch propounds that exceeding the carrying capacity of the physical environment is not the most immediate threat to mankind. Rather, his theory is that the limit to economic growth will first become manifest through social phenomena, not physical pressure on the natural resource base. For example, Hirsch proposes that agricultural production in peri-urban areas will not become threatened or cease because the environment's limits are reached (e.g. because of an excessive load of toxins on the land or because of soil erosion) but because agriculture will succumb to socio-economic forces, 'At the limit, only investment bankers or corporate lawyers can afford to own farms in areas within convenient travelling distance of large cities and rich suburbs.' Another example used by Hirsch is that human health will, in the final analysis, be threatened less by noxious chemicals than by a breakdown in trust between the medical professions and the public which they serve, leading to an increasingly polarised situation characterised by litigation and the concomitant entrenchment of the medical profession.

It is probably true that in many cases, the limits to economic growth will be reached because of social and not environmental forces. However, it is suggested here that there are also real environmental limits to growth, which must be heeded, and that in the search for sustainable development we must be aware of and guard against nearing, reaching or passing these limits, whatever their nature may be. That is, sustainable development is not about refuting the views of Meadows *et al* (1972) with reference to those of Hirsch (1977), but of paying serious attention to both, and trying to reconcile them. Such an approach is inherently worthwhile, although it is also inherently more complex than focusing purely on social, environmental or economic affairs and it is highly unlikely that we can even talk of sustainable development let alone achieve it without adopting such an approach. The inter-linkages between social, economic and environmental spheres are, after all, so widespread that they can rarely be treated in isolation. Indeed, as Cooper (1981) noted in *Economic Evaluation and the Environment*, 'Virtually all acts of production and consumption affect the physical environment in some way or another'.

The realisation of the need to resolve social, economic and environmental problems in a more integrated fashion has continued to grow, and although there is quite some way to go yet, the world of social science is now starting to recognise the rôle which it will have to play in the pursuit of sustainable development. Howard Newby (1990) notes, 'If we are to adapt our life-styles from unsustainable to sustainable patterns of resource utilisation, then the tools of social science analysis are indispensable.' In a later paper (1991) Newby expands on this sentiment, making an impassioned plea:

> ... environmental issues are once more in the forefront of political debate, for in the broadest sense they are deeply political, raising concerns about the expansion of individual choice and the satisfaction of social needs, about individual freedom versus a planned allocation of resources, about distributional justice and the defence of private property rights, and about the impact of science and technology on society. Beneath the concern for 'the environment' there is, therefore, a much deeper conflict involving fundamental issues about the kind of society we wish to create for the future (p.2).

Thus, in summary, a number of the traits of sustainable development emerge from the economic, social and political literature. A fuller description of what it means to develop sustainably will only be possible when the social and economic criteria are combined with those clues to the meaning of sustainable development that have emerged from the literature on the environment. Based on the analysis of the literature reviewed in this chapter, the following sixteen propositions are made. It can be concluded that sustainable development:

1. Recognises that the inter-linkages between social, economic and environmental spheres are so compelling that they can rarely be treated in isolation. One cannot talk of sustainable development without adopting such an approach, let alone achieve it.
2. Rejects large scale industrial production.
3. Involves bringing over-exploitation to an end.
4. Invokes the application of scientific/social scientific knowledge through appropriate technologies, to advance well-being. As great harm can be caused by technology developed or used in an inappropriate manner, sustainable development involves value judgements on its use.
5. Whilst not incompatible with all economic growth, it cannot be achieved at the same time as year-on-year economic growth.
6. Is inextricably linked to the wise use of resources in the pursuit of economic well-being. The over-consumption of resources will add little to

development as a society or to individual quality of life. Waste is to be rejected, not aspired to.
7. Recognises the ultimate dependence of people on the health of all aspects of the environment, and requires protection of the land, sea and atmosphere as well as wholly man-made and modified environments.
8. It is therefore about maintaining the vitality of the environment in order that the vitality of civilisation may be maintained.
9. Has a global dimension. That is, the concept of sustainable development addresses issues on a macro-scale.
10. Is about the ability of different populations to adhere to common principles and aims but to act in a diversity of ways to achieve them, according to local circumstances.
11. Is achieved by putting the principle of subsidiarity into practice, and ensuring that decision-making in all areas of our lives occurs at the lowest feasible level.
12. Is a call to international improvement in the health and wealth of people and their environment, with a special focus on the needs of poor nations.
13. In a global sense will result not only from instruments of foreign policy. The implementation of many domestic policies will also have an impact on the state of the world, and actions undertaken by governments, businesses or individuals should have regard to the effects beyond national boundaries, especially in the Third World.
14. Relies on the *precautionary principle*. Damage caused to the planet and atmosphere by inappropriate development is irreversible on anything other than geological timescales. Where there is doubt over the likely extent of negative impact resulting from an action, it should be avoided until the side-effects can be predicted with greater confidence.
15. Must be achieved gradually and be based on realistic targets. It must also be accompanied by an objective strategy of public education, which explains to the general public the reasons behind setting the targets. Any strategy of public education must be credible to the 'sceptical' not only the 'converted'.
16. Requires the consignment of social prejudice to the dustbin of history.

These dimensions of sustainable development were arrived at through analysis of the socio-economic literature. A study of the environmental and sustainability/sustainable development literature in turn gives rise to a further fourteen defining principles. It is apparent that sustainable development:

17. Requires a modification of the dominant Judeo-Christian value system which believes that God made the world to meet human needs. Although humans are a very special part of the natural world, by believing that they are anything less than a fully integrated part of it we are unlikely to attain such development.
18. Aims to curb population growth, and prescribes that if the human population should continue to expand, ever greater care must be taken in our treatment of the natural environment as it does so.
19. Depends upon the health of the natural goods inherited from earlier generations (e.g. the atmosphere and the Antarctic).
20. Will not be achieved without a widespread acceptance of the view that there are fundamental environmental limits to economic growth (carrying capacity).
21. Requires economies to rely less on fossil fuels as these are non-renewable and their combustion leads to significant atmospheric pollution. Cheaper, more reliable and abundant sources of renewable energy must be developed.
22. Aims to internalise any negative social and environmental externalities (costs) into the economic equation - the polluter pays principle is one example.
23. Requires governments to put in place a system of redress to ensure the internalisation of the costs of environmental damage.
24. Further to the above, will not be achieved in the United Kingdom, without the lead of Parliament and Central Government.
25. Requires that economic and environmental impacts are not addressed in an incremental manner, but that long-time horizons are applied in line with the 'doubling times' advocated by Meadows *et al* (1972, 1992).
26. Considers the well-being of future generations. Inter-generational equity seeks to ensure that future generations will have opportunities equal to today's.
27. Should be based on the principle of equality of opportunity for all. Whilst it is not a socialist doctrine requiring artificial strictures to ensure equality, its realisation depends on a commitment to *social justice* where minimal standards of living are safeguarded for all.
28. Will not be achieved without the provision of sustainable livelihoods through meaningful work which causes no harm to social, environmental or economic systems.

29. Must be achieved in a manner offering value-for-money. To act otherwise would be wasteful and thus incompatible with sustainable development.
30. Is dependent on rural areas playing a more important role in national life.

Conclusion

It can be seen through reviewing the social/economic and environmental sciences literature that many of the conservation aspects of sustainable development were derived from the social sciences (for example the need to view the environment as a global system and the need to use resources wisely). Conversely, many of the socio-economic dimensions were rooted in the environmental literature (the need to protect public goods and ameliorate the effect of negative externalities and the need to ease population growth).

It is important to note that not all aspects of sustainable development have been identified. For example, the need to maintain biodiversity and a generous stock of genetic material on the planet was not drawn out in the definition framed in this chapter, despite the fact that it is a central aim of sustainable development and was viewed as important enough for a legally binding Convention on Biological Diversity to be signed by the majority of countries at the 1992 United Nations Earth Summit in Rio de Janeiro. This and other aspects of sustainable development such as the need to prevent the proliferation of nuclear armaments. the dangers posed to food production by the existence of the 'sunshine limit'. the desire to preserve the culture of indigenous peoples in far-flung corners of the globe. and the steps required to prevent further deterioration of the ozone layer are not listed. In the final analysis, the end result of seeking to define sustainable development is to find that the concept is too large and unwieldy to capture in a generic sense for research, policy development or practical purposes.

Sustainable development does not necessarily lie in the ability to solve multi-faceted economic, environmental and social problems in a precise fashion, nor even the ability to determine exactly the nature of relationships between these three spheres. Our ability to model such complex systems is still relatively undeveloped. Rather, it is the willingness to acknowledge the inevitability of these inter-relationships, and the desire to understand the complex ways in which they are related, which makes the notion of sustainable development more than the sum of its constituent parts.

Annex One: Definitions of Sustainability

A wide variety of definitions of sustainability and sustainable development exists. In choosing the Brundtland definition for sustainable development, it is useful to refer to the twenty-four definitions of sustainable development contained in the Annex to Pearce *et al* (1989). Although many of the key policy documents refer to 'sustainability', they often fail to define it. These documents are listed after the following definitions.

Definitions

1. Porritt (1984) maintains that: 'The radicalism that informs the politics of ecology leads to tough but always logical conclusions.' As an example of one such conclusion he states that 'Sustainability and industrialism are mutually exclusive.'
2. The World Commission on Environment and Development (1987), commonly known as the Brundtland Commission, states: 'A development path that is sustainable in a physical sense could theoretically be pursued even in a rigid social and political setting. But physical sustainability cannot be secured unless development policies pay attention to such considerations as changes in access to resources and in the distribution of costs and benefits. Even the narrow notion of physical sustainability implies a concern for social equity between generations, a concern that must logically be extended to equity within each generation.'
3. Pearce (1989) expounds that: 'The basic rule of sustainability - leaving the same or an improved resource as a bequest to the future - is open to two broad interpretations. Everyone is familiar with the idea of capital wealth - the stock of machinery, factories, roads made by man. There is human capital too - the stock of knowledge that advances as man discovers, experiments and thinks. By now we should also be familiar with the idea of environmental wealth: the stock of natural assets such as tropical forests, freshwater, fisheries and wildlife. Less familiar is the idea of thinking of other environments as wealth too. The ozone layer is an environmental asset, as are the fundamental biogeochemical cycles that regulate the earth and life upon it.'
4. Pearce *et al* (1989) speak of, 'Sustainability as resilience, where the basic system properties for natural populations, communities and ecosystems are therefore *productivity* (in terms of numbers/biomass of individual species), *stability* (constancy) and *resilience* (sustainability)'. Emphasis in original.

5. Dobson (1990) suggests: 'Amid the welter of enthusiasm for lead-free petrol and green consumerism it is often forgotten that the foundation-stone of Green politics is the belief that our finite Earth places limits on industrial growth. This finitude, and the scarcity it implies, is an article of faith for Green ideologues, and it provides the fundamental framework within which any putative picture of a Green society must be drawn. The guiding principle of such a society is that of sustainability, and the stress on finitude and the careful negotiation of Utopia that it seems to demand, forces political ecologists to call into question green consumerist-type strategies for environmental responsibility. In this respect it is principally the limits to growth thesis, and the conclusions drawn from it, that divides light-green from dark-green politics.'
6. Rees (1990) writes, 'True sustainability requires that we recognise the reality of ecological limits to material growth and the need to live on the interest of our remaining ecological capital.'
7. IUCN/UNEP/WWF (1991) pronounced that sustainability is: 'A characteristic of a process or state that can be maintained indefinitely.'
8. In his excellent paper devoted to teasing out the history and meaning of the concept of sustainability, Kidd (1992) concludes: '...while the word sustainability is now widely invoked to justify action or lack of action, there is as yet no consensus on the precise meaning of the term. This is in large part because the term has roots in a number of equally valid strains of thought that are not only widely diverse but also incompatible. This paper is intended to demonstrate that there is not, and should not be, any single definition of sustainability that is more logical and productive than other definitions. The central point of the paper is that those who use the term sustainability should always state precisely what they mean by the term.' (This last sentence is particularly poignant in view of the fact that the key policy documents listed below carry no definition of sustainability.)
9. Meadows et al (1992) defined the criterion of sustainability as being 'that no future society finds itself built around the use of a resource that is suddenly no longer available or affordable.'
10.The Ontario Round Table on Environment and Economy (1992) defined sustainability as: '...anticipation and prevention, efficiency and innovation, quality and respect, commonality and consensus, and links between the environment and the economy and between information and decisions...'
11.The Canadian Round Tables on Environment and Economy (1993) note that sustainability embraces '...the concept that environmental, economic and social needs are complex and require integrated decision making. More

than ever, we understand how decisions made today affect the quality of life for future generations.'

12.The Countryside Commission (1993) has written that: 'The world is constantly changing as natural forces and human activities interact. Sustainability seeks the most benign form of these interactions. It is a dynamic not an absolute state ...' continuing that 'this wider term reflects the fact that not all issues concerned with the environment, the use of resources, and human equity are associated with 'development'. In our view, sustainability implies that human use and enjoyment of the world's natural and cultural resources should not, in overall terms, diminish or destroy them. Different disciplines have contributed their own perspectives to sustainability. the economic and ecological domains are most generally recognised. We believe that the concept also has significant social and cultural dimensions.'

13.Hall and Ingersoll (1993) suggest that: 'Sustainability, particularly sustainable business strategy, is not a definition. It is a process. A process of discovery. A process of learning.'

14.Scottish Natural Heritage (1993) expressed a belief that 'sustainability involves finding ways of maintaining human welfare which do not damage the foundations on which it rests.'

15.Bryden (1995) simply suggests that sustainability 'represents inter-generational equity...'

16.Scottish Natural Heritage (1995) has proposed that 'conservation must become a principle of all social and economic activity. This is the philosophical stance which underlies the concept of sustainability.'

17.Clayton and Radcliffe (1996) explain that sustainability '...means taking steps to try to reduce the risk that environmental and related problems will seriously affect or jeopardise the human species at some future time, and thereby to ensure that future generations have a reasonable prospect of a worthwhile existence... It requires finding ways in which the human species can live on this planet indefinitely without compromising its future.'

Key Documents Offering No Definition of Sustainable Development

The following documents are notable by the fact that they make commitments to sustainability, but fail to define it.

1. British Government Panel on Sustainable Development (1995), *First Report*, Department of Environment, London.

2. Department of the Environment (1993), *UK Strategy for Sustainable Development: Consultation Paper*, Department of Environment, London.
3. Royal Commission on Environmental Pollution (1994), *Eighteenth Report: Transport and the Environment*, HMSO, London.
4. Rural Focus Group (1995), *Rural Framework: A Progress Report*, Scottish Office, Edinburgh.
5. Scottish Enterprise (undated), *The Network Strategy*, Scottish Enterprise, Glasgow.
6. Scottish Office (1992), *Rural Framework*, Scottish Office, Edinburgh.
7. Secretary of State for the Environment *et al* (1994a), *Sustainable Development: The UK Strategy* Cm 2426, HMSO, London.
8. United Nations (1992), *Earth Summit '92: The UN Conference on Environment and Development: Rio de Janeiro 1992*, Regency Press Corporation, London.

Notes

1. The Brundtland definition has been employed by the new Labour Government in its February 1998 document *Sustainable Development: Opportunities for Change* (Consultation Paper on a Revised UK Strategy).

References

Boulding, K.E. (1970), *Economics as a Science*, McGraw-Hill, New York.
British Government Panel on Sustainable Development (1995), *First Report*, Department of Environment, London.
Bryden, D. (1995), *Tourism and Environment: Maintaining the Balance*, paper delivered to *Tourism and Recreation: A Sustainable Approach for Rural Areas: APRS Conference*, Perth.
Canadian Round Tables on Environment and Economy (1993), *Building Consensus for a Sustainable Future: Guiding Principles*, Round Tables on the Environment and Economy in Canada.
Clarke, S.F. and Roome, N.J. (1993), *Towards the Management of Environmentally Sensitive Technology: A Typology of Collaboration*, paper delivered to *Designing the Sustainable Enterprise: The Second International Conference of the Greening of Industry Network*, Cambridge MA.
Clayton and Radcliffe (1996), *Sustainability: A Systems Approach*, Earthscan, London.
Cooper, C. (1981), *Economic Evaluation and the Environment*, Hodder and Stoughton, London.

Countryside Commission (1993), *Sustainability and the English Countryside: Position Statement* CCP 432, Countryside Commission, Cheltenham.

Daly, H.E. (1973), 'How to Stabilise the Economy', in *The Ecologist*, Vol. 3, No. 3, pp. 90-96.

Daly, H.E. and Cobb, J.B. (1990), *For the Common Good: Redirecting the Economy Towards Community, the Environment and a Sustainable Future*, Merlin Press, London.

Department of the Environment (1993), *UK Strategy for Sustainable Development: Consultation Paper*, Department of Environment, London.

Department of the Environment, Transport and the Regions (1998), *Sustainable Development: Opportunities for Change: Consultation Paper on a Revised UK Strategy*, Department of the Environment, Transport and the Regions, London.

Dobson, A. (1990), *Green Political Thought: An Introduction*, HarperCollins Academic, London.

Durning, A.T. (1992), *How Much is Enough?*, Earthscan, London.

Galbraith, J.K. (1958), *The Affluent Society*, Hamish Hamilton, London.

Galbraith, J.K. (1996), *The Good Society: The Humane Agenda*, Sinclair-Stevenson, London.

Gault, R. (1993), *Fixing the Technological Fix: Beyond Right and Wrong and the Technosphere*, paper delivered to *Technology, the Environment and Ethics* conference, Aberdeen.

Goldsmith, E. *et al* (1972), *A Blueprint for Survival*, in *The Ecologist*, Vol. 2, No. 1.

Green Gauge (1996), *Green Gauge '96: Indicators for the UK Environment*, Green Gauge.

Hall, S. and Ingersoll, E. (1993), *Leading the Change: Competitive Advantage from Solution-Oriented Strategies: A Stakeholder-Based Learning Approach*, paper delivered to *Designing the Sustainable Enterprise: The Second International Conference of the Greening of Industry Network*, Cambridge MA.

Hirsch, F. (1977), *Social Limits to Growth*, Routledge and Kegan Paul, London.

Hodson, H.V. (1972), *The Diseconomies of Growth*, Earth Island, London.

Howard, E. (1965), *Garden Cities of To-Morrow*, Faber and Faber, London.

IUCN/UNEP/WWF (1991), *Caring for the Earth: A Strategy for Sustainable Living*, Earthscan.

Kemp, R. (1993), *Technology and the Transition to a Sustainable Economy*, paper delivered to *Designing the Sustainable Enterprise: The Second International Conference of the Greening of Industry Network*, Cambridge MA.

Kidd, C.V. (1992), *The Evolution of Sustainability*, in *Journal of Agricultural and Environmental Ethics*, pp. 1-26.

Kumar, A. (1993), *Sustainable Development: A Strategy for Sustainable Livelihoods*, paper delivered to *Designing the Sustainable Enterprise: The Second International Conference of the Greening of Industry Network*, Cambridge MA.

MacCannell, D. (1976), *The Tourist: A New Theory of the Leisure Class*, Macmillan, London.

Marx, K. and Engels, F. (1965), *The German Ideology*, Lawrence and Wishart, London.

Meadows, D.H., Meadows, D.L., Randers, J. and Behrens, W.W. (1972), *The Limits to Growth: A Report for the Club of Rome's Project on the Predicament of Mankind*, Pan, London.

Meadows, D.H., Meadows, D.L. and Randers, J. (1992), *Beyond the Limits: Global Collapse or a Sustainable Future*, Earthscan, London.

Mishan, E.J. (1967), *The Costs of Economic Growth*, Staples, London.

Newby, H. (1990), *Ecology, Amenity and Society: Social Science and Environmental Change*, in *Town Planning Review*, Vol. 61, No. 1, pp. 3-21.

Newby, H. (1991), *One World, Two Cultures: Sociology and the Environment*, paper delivered to the *Fortieth Anniversary of the Founding of the British Sociological Association*.

Onions, C.T. (ed.) (1973), *The Shorter Oxford English Dictionary: On Historical Principles*, Oxford University Press, Oxford.

Ontario Round Table on Environment and Economy (1992), *Restructuring for Sustainability*, O.R.T.E.E, Toronto.

Orwell, G. (1962), *The Road to Wigan Pier*, Penguin, Harmondsworth,.

Osborn, F. (1969), *Green Belt Cities* 2nd ed., Evelyn, Adams and Mackay, London.

Pearce, D. (1989), *Sustainable Development: An Economic Perspective*, I.I.E.D, London.

Pearce, D., Markandya, A. and Barbier, E. (1989), *Blueprint for a Green Economy: A Report for the UK Department of the Environment*, Earthscan, London.

Pirages, D.C. and Ehrlich, P.R. (1974), *Ark II: Social Response to Environmental Imperatives*, San Francisco, W.H. Freeman and Company.

Porritt, J. (1984), *Seeing Green: The Politics of Ecology Explained*, Basil Blackwell, Oxford.

Redclift, M. (1987), *Sustainable Development: Exploring the Contradictions*, Routledge, London.

Rees, W.E. (1990), *The Ecology of Sustainable Development* in *The Ecologist*, Vol. 20, No. 1, pp. 18-23.

O'Riordan, T. and Cameron, J. (eds.) (1994), *Interpreting the Precautionary Principle*, Earthscan, London.

Rostow, W.W. (1960), *The Stages of Economic Growth: A Non-Communist Manifesto*, Cambridge University Press, London.

Royal Commission on Environmental Pollution (1994), *Eighteenth Report: Transport and the Environment*, HMSO, London.

Rubenstein, D. (1994), *Environmental Accounting for the Sustainable Corporation: Strategies and Techniques*, Quorom, London.

Rural Focus Group (1995), *Rural Framework: A Progress Report*, Scottish Office, Edinburgh.

Schumacher, E.F. (1974), *Small is Beautiful: A Study of Economics as if People Mattered*, Sphere, London.

Scottish Enterprise (undated), *The Network Strategy*, Scottish Enterprise, Glasgow.

Scottish Natural Heritage (1993), *Sustainable Development and the Natural Heritage: The SNH Approach*, Scottish Natural Heritage, Battleby.

Scottish Natural Heritage (1995), *The Natural Heritage of Scotland: An Overview*, Scottish Natural Heritage, Battleby.

Scottish Office (1992), *Rural Framework*, Scottish Office, Edinburgh.

Secretary of State for Economic Affairs (1965), *The National Plan* Cmnd 2764, HMSO, London.

Secretary of State for the Environment *et al* (1994a), *Sustainable Development: The UK Strategy* Cm 2426, HMSO, London.

Secretary of State for the Environment *et al* (1994b), *Biodiversity: The UK Action Plan* Cm 2428, HMSO, London.

Skirbekk, G. (ed.) (1994), *The Notion of Sustainability and its Normative Implications*, Scandinavian University Press, Oslo.

Smith, A. (1982), *The Wealth of Nations: Books I-III*, Penguin, Harmondsworth.

Tainter, J.A. (1988), *The Collapse of Complex Societies*, Cambridge University Press, Cambridge.

United Nations (1992), *Earth Summit '92: The UN Conference on Environment and Development: Rio de Janeiro 1992*, Regency Press Corporation, London.

Veblen, T. (1899), *The Theory of the Leisure Class: An Economic Study of Institutions*, Macmillan, London.

World Commission on Environment and Development (1987), *Our Common Future*, Oxford University Press, Oxford.

3 Environmental Beliefs and Behaviour in Scotland

JAMES McCORMICK and ELEANOR McDOWELL

Introduction

Survey research has become a more common means of measuring the 'pulse' of contemporary society in recent years. The purpose of this chapter is to explore how opinion surveys are used to measure environmental beliefs and behaviour in Scotland. Policy-makers and environmentalists are far from agreed on which reforms can best promote sustainable development. As devolution unfolds, more data will be required if the new Parliament is to do a more effective job than Westminster in shaping appropriate environmental policies for Scotland. Yet the goal is broader than a 'technical fix' based on devising better sources of data for decision-makers. Citizen participation will become increasingly important: as voters, service users and consumers in making the behavioural changes which sustainability requires; and in voicing their opinions on existing problems and possible solutions.

The *quality* of participation matters as much as the number of people who have an opportunity to express their views. Local government and health authorities have been at the forefront of using more deliberative methods of consultation with the public like citizens' juries. Where decision-makers are genuinely interested in finding out the views of ordinary members of the public, such methods can generate unique insights which cannot be captured by measuring public opinion through snapshot surveys. Nevertheless, opinion polling continues to have a value by providing information on the immediate reaction of a cross-section of the public to particular issues.

As part of the *Environment Scotland* project on which this book is based, a number of questions were placed on a monthly omnibus survey of Scottish public opinion in March 1998. The survey measured attitudes on a number of key environmental policy options (reduced car use, renewable energy and green belt development), the extent of green behaviour (recycling and buying environmentally-friendly goods for example) and awareness of 'sustainable development'. This chapter presents some of the findings of previous opinion research which allows us to compare Scotland with the rest of Britain where appropriate, as well as the results of preliminary analysis of the 1998 Scottish survey.

Measuring Public Opinion

At one level, the concepts organised around sustainable development are uncontroversial. Few of us favour resource depletion which pays no regard to conservation, or are prepared to say that what happens in our own lifetime is all that matters. Sustainable development is nothing if not based on some notion of justice between the generations. So it should come as no surprise to find that most of the polling evidence in recent years points to substantial majorities of the Scottish and British public taking up pro-environment positions against waste, pollution and congestion. Before reviewing some of the evidence, some of the criticisms, which are often levelled against opinion polling data, are discussed.

First it is argued that people have a tendency to say one thing and do another. The most common example prior to the 1997 General Election was the distinction between political attitudes, voting intentions and voting behaviour. During the four periods of Conservative government since 1979, there was no conversion towards the central tenets of Thatcherism. In fact there was a marked swing in favour of broadly collectivist solutions. By the end of the 1980s, the British Social Attitudes survey recorded a growing majority in favour of more spending on public education and the NHS, and higher taxes in order to pay for it. Parties of the centre-left never lost the battle of attitudes. A dwindling minority supported the philosophical Conservative position -lower taxes and lower state spending -. Yet the Conservative share of the vote in Britain as a whole was virtually unchanged throughout the period. Why could the public not be trusted to hold its nerve in the polling booth?

The distinction between attitudes and behaviour is more subtle than first appears. Consistently 60 per cent of voters said they favoured better public services over tax cuts, and just under 60 per cent of those who voted in General Elections between 1979 and 1992 supported parties which proposed 'tax and spend' packages. The figures were closer to 75 per cent in Scotland. The problem, according to this view, was the distorted picture of the electorate's wishes created by the first-past-the-post voting system at Westminster. An alternative view is that Scots also took advantage of Conservative reforms like council house sales and share ownership even if they failed to give the party credit through the ballot box - in other words, neither core values nor electoral behaviour may have changed, but consumer behaviour did.

Second, there is the charge that polling is an increasingly blunt instrument for measuring the subtleties and assumed volatility in public opinion. How far this is accurate depends on what question is being asked and whether a poll is tapping into a well-established public dialogue or catching respondents 'cold' (and poorly informed). On the constitutional future of Scotland, there was remarkable consistency in preferences in the last decade despite one or two high-profile 'rogue' polls in the years before the devolution referendum. Perhaps most telling was the very small number of people who were uncertain about how they would vote in September 1997. On the first question, whether there should be a Scottish Parliament, the proportion was no greater than the numbers who typically say they do not know how they will vote in a general election (or refuse to answer), and it was significantly lower than in Wales.

Large majorities consistently express disquiet about environmental problems like Dounreay when given the chance to voice their opinions. However we cannot assume that snapshot measurement of concerns which may ebb and flow as media coverage moves on means there is a ready-made constituency for an enlightened Scottish Parliament to tap into. What we conclude about polling depends on the specific issues being explored and how the questions are posed. The difference between the breadth and depth of attitudes, their consistency and resilience, and the links between attitudes and behaviour, are now discussed.

Breadth and depth

Attitudes are commonly measured by trade-off prompts or by priority-rankings. The critics of a Scottish Parliament did not dispute the consistent finding that 70 per cent of Scots supported the devolution before and during the referendum campaign. Instead they argued that it was not a prominent issue in general election campaigns. On a list of most important concerns, the constitutional question typically ranked seventh or eighth, behind jobs, education, health and crime. In a trade-off question posed in 1994, two-thirds of Scots believed that environmental protection was more important than either economic growth or holding down prices (McCaig and Henderson, 1995), yet the environment barely registers among self-reported Scottish election priorities. One conclusion is that the public is prepared to give the 'correct' answer in an opinion poll without necessarily having an impact on their own set of priorities. Another British study stated that ‘virtually everybody expresses some disquiet about environmental issues’ but that concern is broader than it is deep-rooted (Witherspoon and Martin, 1993).

The relatively weak expression of green concerns among voters may encourage some decision-makers to doubt the importance of action. They may prefer to wait until pro-environment policies are proved necessary beyond reasonable doubt or until the electorate decides it is time to change priorities. Such a model of decision-making is partial at best. It overlooks the other points of pressure on national governments, ranging from issue-based campaigning by the environment movement to EU legislation and international agreements. This model also tends to commit the error of compartmentalising policies as if they were wholly distinct, assuming because 'environment' does not have the same immediate salience to voters as 'unemployment' or 'the National Health Service' it must be of little concern. Just as the new Parliament ought to be a means of addressing the policy priorities of Scots and not an end in itself, a sustainable environment should contribute to progress on jobs and health (as well as result from it). If these links remain weak it reflects a general failure to articulate them in ways that most people can relate to. If the penetration of green beliefs and behaviour is indeed 'both shallow and patchy' (Witherspoon and Martin, 1993) that is a reason for action rather than a sign of irrelevance.

Consistency and resilience

Unfortunately there is little scope for tracking Scottish attitudes on the environment over time. This has been recognised by the inclusion of a number of questions on the 1998 survey for comparison with earlier results. Even if a more comprehensive set of time-series data existed, there would still be strong grounds for caution in drawing conclusions. The precise reasons for attitude formation and related behaviour are inherently difficult to determine. People have conflicting interests and loyalties. Beliefs may persist because they have not been challenged. Where 'cost-free' attitudes are challenged with practical implications, and are reinforced rather than weakened, they can be thought of as *resilient*. Volatility over time in public priorities is likely to reflect the influence of significant events. The general rule that unemployment, health and education are ranked highest was breached (spectacularly so) by the poll tax. Environmental concerns have grown in importance following the Chernobyl disaster, oil tanker spillages and the 'flash-flood' success of the Green Party in 1989. Where environmental concerns are among the highest priorities in particular localities, they usually reflect the threat of environmental deterioration (such as a new road or factory being built on green belt land, or a waste incinerator being located in a densely-populated area).

Defining Green Beliefs

In this section, the aggregate findings of the 1998 Scottish study are discussed alongside the results of earlier studies of environmental attitudes, or green beliefs. The distribution of attitudes is then considered along a number of key explanatory variables.

There has been an increase in general awareness of environmental problems like the threat to the ozone layer, resource depletion and the Greenhouse Effect. Scottish attitudes appear to have followed a similar trend as in the rest of Britain. A study of attitudes commissioned by the Scottish Office (Wilkinson and Waterton 1991) revealed that the Scottish public were concerned most about (in order) the pollution of rivers, lochs and seas and the dumping of raw sewage at sea, followed by the quality of drinking water, nuclear waste and damage to the ozone layer. Comparing Scottish attitudes with England and Wales revealed a broadly similar order of priorities, the top two issues being the same.

According to Witherspoon and Martin's 1993 study however, Scottish respondents were less concerned than in the rest of Britain about 'global green issues' and pollution, perhaps because concern over jobs and growth has tended to be pitted against environmental protection. Respondents in the south of England were the most concerned, even when geographical differences in employment, earnings, social class and demography have been taken into account. On the basis of this study at least, Scots appeared to be relatively less concerned *per se*, not because those factors which are otherwise associated with less concern are more common in Scotland. However, Scots were clearly *not* less concerned than the average about nuclear power.

The nuclear issue remains controversial, reflecting concerns over security at Dounreay as well as the concentration of nuclear military sites at Rosyth, Holy Loch and Faslane. The low population density and extensive coastline in the remote north of Scotland offered government planners an ideal location for a nuclear reactor and reprocessing site. Nuclear development at Dounreay, as well as Torness and Hunterston, prompted a series of campaigns by anti-nuclear groups throughout Scotland, arguably maintaining a higher public profile for the issues over a longer period of time than in the rest of Britain.

The 1998 Scottish study does not enable us to take these comparisons within Britain any further, but it does allow attitudinal differences *within* Scotland to be considered (using a standard regional variable, which provides large enough samples in West and East central Scotland for comparative purposes). It would be possible to conclude that Scottish attitudes are a paler shade of green than in the rest of Britain. However the datasets remain too patchy to assume that any such differences are consistent over time or apply to all environmental issues equally.

The studies reviewed here presented respondents with a series of environmental scales ('Green Global', 'Pollution' and 'Nuclear' in Witherspoon and Martin's 1993 study of Britain) and sustainable development scenarios (in McCaig and Henderson's 1995 Scottish study). In the British study, a cluster of concerns about pollution was assessed as a much more serious threat than nuclear or global environment problems. Both studies found around one in ten responses to various questions could be described as the most 'environmentally-correct.' For example, just over one in ten people strongly agreed that more should be done to protect the environment even if it meant higher prices, higher taxes and, controversially, fewer jobs.

What do Scots think? A similarly 'costed' question about energy use was included in the 1998 study. The following proposition was made:

> Scotland should develop alternative sources of energy such as wind and water power, and reduce its use of North Sea oil and nuclear power, even if it costs more in the short term

Around two-thirds of respondents agreed and 21 per cent strongly agreed. Despite the possible cost implications and the reference to Scotland's oil (which continues to have be strong potential as a political symbol), only one in seven respondents disagreed. If this finding is an accurate reflection of public opinion, it suggests clear potential to revisit the policy debate about energy sources more than twenty years on from disputes over North Sea Oil and the 'Save It' campaign for energy conservation. Holyrood's new politicians may have greater incentives to take action around the policy agenda mapped out by Gloyne and Hutton in Chapter Ten than they first realise, despite the broad powers to set energy policy being reserved to Westminster.

The McCaig and Henderson study for the Scottish Office included a number of priority-ranking questions. Respondents were asked to choose the most important features of 'a good quality of life.' More than half chose high levels of employment and good health care. Around one in five chose 'a good quality environment,' slightly more than chose the broadly-defined option of 'economic growth.' In a straight choice trade-off question, two-thirds of Scots said that environmental protection was more important than either economic growth or keeping prices down. These results provide a useful snapshot of responses to the environment versus economy question, but there is also a need to know how people respond to rather more specific questions.

McCaig and Henderson found that Scots were less inclined to have pro-environment attitudes and more likely to emphasise economic and financial concerns when the issue in question is 'closer to home.' Thus there was considerable support for addressing the problems of over-population in developing countries and over-fishing, but opinions were more mixed when the consequences were more immediate. One of their four sustainable development scenarios concerned the pressures on green belt land near the city from supermarket developers. Despite more traffic and loss of green belt, should the demand for more retail space be met by such development? Scots appear to be opposed to tighter regulations on shopping development and restrictions on how people shop. Over 60 per cent agreed that 'it would be wrong to prevent out-of-town shopping if that's what people want' and over 80 per cent agreed that 'if people want to go shopping in their cars it's up to them.' However more than 60 per cent *also* agreed that 'If we carry on like this there won't be much countryside left for our children' and just under half agreed that 'Such developments shouldn't be allowed because they encourage people to use their cars more.' A minority hold what appear to be conflicting views, reflecting the difficulty of capturing the combination of individual beliefs about difficult environmental issues.

One question was placed on the 1998 Omnibus Survey to test responses to a similar scenario on industrial development in the green belt (an issue which local government and the Local Enterprise Companies are familiar with). The following proposition was made:

> An international company has announced it wishes to build a new factory in Scotland. It wishes to locate in a 'greenfield' site on the outskirts of town, while the Council and environmental groups want it to be built on derelict land near to the main road. The company says if it cannot locate in the green belt it will move elsewhere. Do you agree or disagree with the following:
>
> At the end of the day, extra jobs are more important than protecting the countryside.
>
> If we carry on like this, there won't be any countryside left in the future.

Respondents were evenly divided, splitting 41 per cent against more jobs if this was the price and 39 per cent in favour. One in seven respondents either strongly agreed or strongly disagreed. In contrast almost seven out of ten respondents agreed with the second statement, with one in four strongly agreeing that there are grounds for concern. This is marginally (4 per cent) above the figure in the study by McCaig and Henderson (1995). This suggests that a minority of Scots favour the short-term advantages even though they are aware of, and concerned about, the medium/longer-term disadvantages.

It is worth noting here that as many as one in five respondents neither agreed nor disagreed with the first statement. This may reflect unease about having to make a difficult tradeoff. Although the proposition suggests that extra jobs in the greenbelt carries an environmental price tag, sustainable development means that economic progress and environmental protection should reinforce each other. Public opinion is clearly divided. It is the responsibility of politicians and other economic decision-makers to reduce the need for such trade-off situations in the future.

Drawing together the existing Scottish and British evidence on environmental beliefs, how do they relate to people's different backgrounds? Another set of questions was included in the 1998 Omnibus to measure environmental attitudes. Table 3.1 shows the level of support for a number of options designed to persuade people to reduce car use. The options are ranked according to the proportion that agreed that each option should be used for that purpose.

Table 3.1 Respondents agreeing that these methods should be used 'to persuade people to use their cars less often, in order to reduce pollution and congestion'

Reduce the cost of using buses	69%
Reduce the cost of using trains	53%
More *reliable* transport service	52%
More public transport after 6pm	46%
Offer more Park and Ride facilities	36%
Create more cycle routes	23%
Create more bus lanes	20%
Introduce a toll if drivers wish to drive into the city centre	13%
Raise the price of petrol in cities	2%

It is unclear how different people interpreted the question. Some respondents may have personally favoured an option but doubted its effectiveness in reducing car use. Predictably, those measures which would make the environmentally sustainable options easier and cheaper receive much more support than the two options which would raise the cost of urban motoring. In July 1998, the Government's long-awaited White Paper on Transport focused on making integrated transport services a reality. While it offered the 'carrots' of better national information, reduced fares for the retired and safer routes for children to travel to school, it also proposed the 'sticks' of charging for non-residential car parking and city centre tolls. The Scottish White Paper announced the possibility of introducing tolls on the M8 motorway between Glasgow and Edinburgh. Governments have been willing to take unpopular measures such as introducing a petrol price escalator to raise revenue and reduce car use. These initial results, coming ahead of the high-profile debate sparked by the White Paper, suggest that the long-run advantages of new pricing mechanisms will have to be clearly explained if public support is to be secured.

The distribution of attitudes

Of considerably more interest is the pattern of attitudes lying beneath the surface. The results for these questions on reduced car use are assessed along a number of individual and household variables. Table 3.2 presents *the range in responses* (in percentage terms) for each variable. For example it shows on the question of reducing bus fares a 21 per cent gap in support between the most favourable age group and the least favourable, compared with a 5 per cent gap between the sexes. Results for the propositions on energy sources and the jobs/countryside trade-off are also reported in Table 3.2. Responses to a final question measuring awareness of the concept of 'sustainable development' are included in the table.

Age

Younger respondents are consistently more in favour of strategies to reduce car use and older respondents least in favour. Older respondents were also found to express less than average environmental concern by Witherspoon (1995). However the age groups are united in their support for the most popular options of cutting bus and train fares. One unexpected finding is the distinctive set of attitudes among the 55 to 64 age group: those respondents are least in favour of reducing fares (skewing the results in Table 3.2) or providing more public transport in the evenings, and most in favour of providing more park and ride facilities. Older respondents were also least in favour of developing alternative energy sources if it ended up costing more, although the difference was not as marked as on some of the other questions, and it is worth noting that a clear majority was in favour in all age groups.

Respondents aged 45 to 54 were alone in supporting the trade-off in favour of jobs rather than the countryside (by 10 per cent). There was broad concern across all age groups about the future threat to the countryside, with 25 to 34 year olds most likely to agree with the proposition. Although the variation in response to this question were not large (11 per cent), age proves to be the strongest discriminating factor (along with socio-economic group). Finally there was a significant difference in awareness of the term 'sustainable development' between the age groups. While about one in five identified with the term overall, this ranged from one in three of the 25 to 34 year olds, to one in seven of the retired.

Socio-economic group (SEG)

Socio-economic group was, generally, a less powerful means of distinguishing between respondents' attitudes than their age group. The C1s (skilled non-manual workers) emerge as most in favour of the majority of options. A comparison of ABs (professional and managerial workers) with DEs (unskilled manual and economically inactive) produces some surprising findings, not least that professionals are significantly more in favour of reducing train fares and to a lesser degree bus fares, and more reliable public transport, than those at the lower end of the economic ladder. This may reflect differential use of (and probably access to) train services rather than any greater support *in principle* for cutting the cost of public transport.

C1s are also most in favour of finding new sources of energy, although all social groups provide two-thirds support for this option. On the jobs trade-off question, DEs are most firmly in favour of priority for jobs (by 10 per cent) and C1s least so (by 16 per cent), although in no category are half the respondents in favour of either option, largely because as many as one in five respondents did choose one option over the other. Perhaps surprisingly then, DEs are more likely to agree that 'there will be no countryside left' than are ABs, while C1s again stand out as the most concerned. A straight split emerges around awareness of sustainable development: ABs and C1s are 20 per cent more likely than C2s (skilled manual workers) and DEs to identify the term.

Comparing these findings with McCaig and Henderson, Scots in social groups A and B were significantly more likely than DEs to choose 'a good quality environment' from a list of eleven quality of life features (34 per cent to 15 per cent) as well as 'high quality education' (46 per cent to 31 per cent), but less likely to choose 'supply of good quality housing' (15 per cent to 32 per cent). On an aggregate scale constructed from McCaig and Henderson's four scenarios, ABs were more strongly pro-environment rather than pro-economic development, and perhaps more surprisingly they were also the most strongly in favour of government intervention to protect the environment. For example, DEs were least likely to take a regulatory stance against out-of-town shopping development and car use despite having the lowest levels of car ownership. This may partly reflect offended aspirations ('I may not have a car now but if I did I would use it as much as I liked)' in the way that many low-paid workers have reacted against tax increases on the higher-paid which may actually have been to their advantage ('I may not earn enough now to pay a higher rate of tax, but I aim to earn that much one day').

Table 3.2 Green Attitudes in Scotland:
Range of responses (percentage) by survey variables

	AGE	SEG	SEX	HSNG	CAR	EDUC
Reduce car use by......						
Lower bus fares	21	8	5	11	5	15
Lower train fares	19	16	4	8	16	15
More reliable public transport	18	17	10	2	12	12
More evening public transport	21	14	1	5	7	14
Park and ride	26	15	6	9	11	19
Cycle routes	14	14	2	3	0	33
Bus Lanes	11	10	3	2	0	17
City tolls for drivers	10	8	2	4	2	18
Raise petrol prices in cities	4	2	0	1	1	6
'Alternative energy sources should be developed'	16	6	12	0	9	11
'Jobs more important than countryside'	13	13	0	3	9	14
'Risk of no countryside left'	11	11	1	2	6	3
Awareness of sustainable development	18	22	11	6	6	40

Previous studies have found a more complicated pattern of attitudes related to social class in Britain. As with the 1998 Scottish findings, professional and managerial respondents were generally found to have weaker green attitudes than middle/junior non-manual workers (Witherspoon and Martin, 1993), and an alternative analysis by household income showed environmental concern was lower among the better-off (Witherspoon, 1995). However regression analyses to measure the independent effect of each variable have suggested that *educational attainment* explains the distribution of environmental attitudes more accurately than occupational status or household income do on their own. Given these mixed signals, the relationship between Scottish attitudes and education is considered in more detail below.

Sex

Scottish men and women do not appear to hold significantly different views on the environment. Sex turns out to have one of the 'flattest' attitude profiles. Women are rather less likely to favour alternative energy if it costs more (although still six in ten still do) and less aware of the term 'sustainable development.' A surprising finding is that women are 10 per cent less likely than men to favour more reliable transport services as a means of reducing car use. This is presumably a reflection of women's lower rates of car ownership than any lesser reliance on public transport.

Housing tenure (HSNG)

Housing tenure is often assumed to be a resilient determinant of Scottish attitudes. However, the straight split between homeowners (57 per cent of the sample) and the combined category of renters (43 per cent) turns out to be less significant than expected. The differences on how best to reduce car use are modest, although homeowners are rather more in favour of cutting the cost of public transport than are renters. On questions about jobs and the countryside, there are no clear differences. The same is true of attitudes to alternative energy sources, which in itself is surprising. Scottish households who rent typically pay a higher proportion of their incomes in fuel bills and might have been expected to be deterred by the reference to potentially higher costs of developing new sources of energy.

Number of cars (CAR)

Those households without a car, almost one-third of the sample, are now compared with those who have two or more cars (who account for one in seven respondents). 'Car rich' households, defined as having two or more cars, are no less in favour of more bus lanes and cycle routes than the 'car poor', even though these may be considered an inconvenience by some motorists. Surprisingly, the 'car rich' are more likely to support reduced fares and more reliable public transport than the 'car poor'. There is overwhelming opposition to raising urban petrol prices: car ownership does not distinguish attitudes. The 'car rich' are more in favour of alternative energy sources and more likely to be concerned about the loss of countryside than the 'car poor' (who split 38 to 35 per cent in favour of extra jobs, compared to the 'car rich' who split 60 per cent to 29 per cent against).

Educational attainment (EDUC)

Educational attainment emerges as the most powerful factor in discriminating between attitudes. On almost every option for reducing car use, those with a university education are by far the most in favour. The difference between those with a university degree and those who completed their education when they left secondary school is greatest on support for cycle routes, bus lanes and user tolls for city motorists. University graduates are twice as likely as the average to support these options, although the strongest support for any single option comes from those who entered Further Education: three-quarters of them favoured cutting bus fares. This pattern is only reversed for the park and ride option, where university graduates are least in favour.

Support for alternative energy sources also rises with educational attainment, although a wider range of responses is found between the age groups. The jobs/countryside trade-off moves in the opposite direction. Those educated to secondary school level are marginally in favour of extra jobs (42 to 38 per cent) while those who attended a training establishment or vocational college split 54 to 28 per cent against the proposition. The one question, which attracted strong support from each group, was on the future threat to the countryside: two-thirds of respondents agreed with the proposition irrespective of their level of education. The single largest single gap between respondents is in their awareness of sustainable development. More than half of university graduates said they were aware of what it means (53 per cent), four times higher than the level of recognition among those educated to secondary school level (13 per cent).

Although McCaig and Henderson's study did not include educational attainment as a background variable, Witherspoon and Martin also found that education was strongly related to environmental concern, but that the strength of the link depended on the specific question being explored. While the more highly educated were clearly more concerned than others about 'global green issues', they were less concerned about pollution and much less worried about nuclear power than those with the least formal education.

Region

The 1998 survey allows the results to be broken down into four regions (North, East, West and South Scotland), offering a comparison of environmental beliefs and behaviour *within* Scotland. The discussion is limited to West and East Scotland at this stage, since these two regions account for 78 per cent of all respondents and the samples in each are large enough to draw conclusions from. Results are reported rather than shown in Table 3.2 for reasons of space. On all but one attitudinal question, the East of Scotland emerges as more pro-environment than the West. The gap is largest for the jobs-greenbelt scenario, (where respondents from the East are 8 per cent less likely to favour extra jobs at the expense of the greenbelt and 9 per cent more likely to believe that the countryside is at risk); for awareness of sustainable development (12 per cent higher in the East); and most significantly in support of more cycle routes as a means of reducing car use (15 per cent more in favour), probably reflecting Edinburgh's greater number of journeys to work by bicycle compared with Glasgow and the capital city's more active approach to cycling and other alternatives to the car.

Is there a typology of green attitudes?

Is there a set of green attitudes which people tend to endorse or reject on principle? Or do beliefs depend on particular issues, where people may be more likely to adopt a green stance on some questions (perhaps global concerns) than others (local trade-offs). Witherspoon and Martin found relatively high correlations across different attitude scales, although concern about nuclear power was much more strongly correlated with concern about nuclear weapons than with any other environmental issue. On the other hand, McCaig and Henderson's sustainable development scenarios revealed little evidence of Scots who are *consistently* pro-environment or pro-economic development on all issues. They concluded instead that respondents were influenced by the implications of particular statements rather than guided by a general map of principles.

The preliminary results of analysing the 1998 Scottish data are presented here. More detailed statistical testing of the correlations between the variables will be conducted in a follow-up project. The simple method of identifying responses to each question which are at least 10 per cent higher than the sample average is used (the definition of 'significance'). Those characteristics associated with stronger environmental attitudes, ranked by the number of 'significant' responses to the thirteen questions shown in Table 3.2, are:

1. University education	6
2. Two or more cars in the household	5
3. Technical/Vocational College education	5
4. Social Group C1	4

No other characteristic scored significantly above average on more than two questions. These results suggest that there is no clearly identifiable group of people with consistently green beliefs, although a combination of the features shown above (such as households with a university graduate and two cars) is likely to be associated with the strongest pro-environment attitudes.

Defining Green Behaviour

While most people express some concern about environmental problems, regular green *behaviour* and willingness to pay for it are less common. Witherspoon and Martin (1993) conclude that knowledge of an individual's background appears to explain more about their behaviour than their attitudes, especially on their green consumer scale. They presented a hierarchy of green activities from the most commonly done (55 per cent bought environmentally-friendly aerosols) to the least common (17 per cent refused unnecessary packaging and 8 per cent bought organic fruit and vegetables).

What might be termed 'green civic action' was not explored in the studies reviewed here, but represents another form of pro-environment behaviour, often prompted by an environmental threat to a locality. The M73 extension through southern Glasgow is a recent example of road protest; the eight-year campaign against the location of a waste incinerator in Renfrew on the south bank of the Clyde is an example of popular action by an entire community (see Chapter Five in this volume). Few of the people in Renfrew who took part in the campaign would think of themselves as environmental activists. It is doubtful that it had a lasting effect on how people behave as consumers or voters, but it does demonstrates the capacity of ordinary people to organise against an environmental threat. Although green civic behaviour is voluntary, many of those who take part feel they have no alternative to public protest.

Raising awareness further will not be enough to achieve the changes some policy-makers believe to be necessary and urgent. Green behaviour may also be encouraged through regulation and pricing mechanisms. Where charges are raised, consumers are not *required* to change their behaviour. Instead the pattern of incentives is altered to persuade people to make changes in their volume and type of consumption or face higher costs. The annual petrol price escalator is one example of government taking an unpopular action to raise revenue and encourage reduced car use.

If Scots appeared to be less concerned about some environmental problems than the average for Britain, were they also less green as consumers and as taxpayers? Controlling for education, social class and party loyalty (a factor not included in the 1998 study), the Scots and the Welsh were found to be less likely to act as green consumers than the British average (Witherspoon and

Martin, 1993). But they were just as likely to support higher taxes/charges to pay for environmental protection as respondents in the rest of Britain. This contrasts with a Scottish Office study which found that Scots were relatively *more* likely to 'act green' by using ozone-friendly aerosols, picking up litter and so on (Wilkinson and Waterton, 1991). In addition, a more recent market research survey by Mintel (Harrington, 1997) found that Scottish shoppers lead the rest of Britain in buying organic foods. Three out of four respondents were found to be 'sympathetic' to buying organic, although ability to pay the higher costs was identified as a barrier for some households.

A further set of questions was included in the 1998 survey to measure Scottish environmental behaviour. Respondents were asked 'What activities do you undertake to benefit the environment?' Table 3.3 shows the level of activity across nine specified options. They are ranked according to the proportion who undertake them. In line with earlier studies (Witherspoon, 1995 for example), recycling behaviour is the most commonly undertaken activity by some distance. However, that none of the activities are undertaken by half of respondents might be considered disappointing. The numbers who say they engage in green behaviour are generally below those found in earlier studies of Britain (Witherspoon and Martin, 1993; Witherspoon, 1995). Part of the explanation may lie in how the survey questions were posed. In the 1998 study, respondents were invited to judge for themselves whether they were involved in recycling or green shopping, regardless of how regularly they did so, while other studies have recorded higher rates of activity by asking how often respondents acted green.

The evidence on Scottish green behaviour relative to the rest of Britain is mixed. The findings appear to be sensitive to how respondents are questioned and when the studies were conducted, suggesting that the differences are far from consistent. Unfortunately there is a lack of comparable data for us to draw any firmer conclusions on these trends over time.

The Distribution of Greener Behaviour

The raw figures do not tell us much about the pattern of Scottish environmental behaviour. The results in Table 3.3 are measured across the variables used in the analysis of attitudes. Table 3.4 presents the *range* of responses for each variable. It shows, for example, that the difference in recycling behaviour across the age groups is 14 per cent. What can be deduced about green behaviour when we scratch the surface of these aggregate findings?

Age

Contrary to expectations, the relationship between green behaviour and age is not straightforward. On seven of the nine options, youngest respondents are *less likely* than the oldest to be involved, although the differences are not significant with the exception of saving energy (which partly reflects the different stages of household formation). Young respondents are rather more likely than the oldest to choose cosmetics which have not been tested on animals and to buy ozone-friendly goods. Yet the most marked differences are not between these ends of the age spectrum. On buying cosmetics for example, the widest gap is between twenty-somethings and forty-somethings. The 25-34 and 35-44 age groups are least likely and the 45-54 year olds most likely to recycle paper (a gap of 10 per cent). But the 45-54 age group is the least likely of all to recycle bottles and cans. In only one category and for one option does a majority of respondents engage in green behaviour: 55 per cent of people aged 35-44 recycle bottles and cans.

Table 3.3 Which of these activities do you undertake which you consider to benefit the environment?

Recycle bottles and cans	48%
Recycle paper	40%
Use unleaded petrol	33%
Buy environmentally/ozone friendly products	28%
Pick up litter	25%
Buy cosmetics/toiletries not tested on animals	22%
Save energy	22%
Cut down on car use	13%
Buy organic fruit/vegetables	9%
Other mention	16%

Socio-economic group (SEG)

The pattern of environmental behaviour is more consistently related to socio-economic group than to age. On six of the nine options, ABs are more likely than DEs to 'do the green thing', particularly so for recycling, using unleaded petrol and cutting car use (the latter two clearly reflecting different levels of car ownership). However there is virtually no difference between them when it comes to buying organic foods. Only 8 per cent of ABs buy organic fruit and vegetables compared to 10 per cent of DEs who have fewer resources and less scope to exercise choice. On the other main consumer option, buying environmentally-friendly goods, there is little difference between social groups although it is worth noting that DEs appear rather *more* likely to 'buy green' than ABs (30 per cent to 25 per cent). After controlling for educational attainment (the strongest influence on behaviour in Witherspoon and Martin's study), the intermediate occupational categories covering middle and junior-ranking non-manual workers emerge as more likely than either professional or manual workers to be green consumers.

Table 3.4 Green Behaviour in Scotland:
Range of responses (percentage) by survey variables

	AGE	SEG	SEX	HSNG	CAR	EDUC
Activities undertaken to benefit the environment.....						
Recycle bottles	14	17	0	10	22	17
Recycle papers	10	15	0	7	20	22
Unleaded petrol	17	17	9	13	(45)	30
Env/ozone friendly	14	6	7	3	5	14
Pick up litter	10	11	3	1	2	5
Cosmetics	16	7	10	0	3	15
Save energy	11	13	8	2	11	22
Reduced car use	7	8	7	4	4	23
Organic foods	3	2	4	0	2	21

Sex

Someone's sex is a relatively weak predictor of green behaviour. There are no reported differences in household recycling. The greater use of unleaded petrol and reduced car use among men are almost wholly a reflection of uneven car ownership. The only significant difference between the sexes applies to shopping. Women are rather more likely than men to buy environmentally-friendly goods and toiletries that have not been tested on animals, although fewer than one-third do either.

Housing tenure (HSNG)

Housing tenure is also a relatively weak indicator of behaviour, as it is for attitudes. Homeowners are rather more likely to recycle household goods and use unleaded petrol than those who rent their home, but there are virtually no differences between them on the other consumer options. Those who pay a mortgage are no more and no less likely than those who pay rent to be green shoppers.

Number of cars in household (CAR)

Car ownership strongly distinguishes respondents on both recycling options. A clear majority of the 'car rich' recycle bottles and cans (62 per cent) and paper (52 per cent). They are at least one fifth more likely than the 'car poor' to do so, reflecting their greater opportunity to recycle on a regular basis. The association between car ownership and greener attitudes extends to this form of behaviour, although the irony of making additional car journeys in order to recycle should not be lost on policy-makers. The very low number of car owners who have cut down on car use (only one in seven) and who use unleaded petrol (still below half) are striking. Government is committed to reducing car use through a combination of higher petrol prices, road pricing, charging for non-residential car parking and improving public transport. These figures indicate the scale of the challenge for politicians. Nor do the 'car rich' make particularly green shoppers: if anything they are marginally less likely to buy environmentally-friendly goods.

Educational attainment (EDUC)

Education is the strongest indicator of green behaviour as well as attitudes. University graduates are more likely than secondary school graduates to act green on seven of the nine options. They are much more likely to undertake *the least common options* of reducing car use, saving energy and buying organic foods, as well as the more popular option of paper recycling. These findings are consistent with other studies which highlight the clear association between more education and green consumer behaviour, perhaps because education raises awareness and improves ability to identify pro-environment behaviour (Witherspoon and Martin, 1993; Witherspoon, 1995).

Region

On all but two of the options mentioned in the survey, respondents living in the East of Scotland are more actively green than those in the West (the exceptions being for unleaded petrol and buying environmentally/ozone friendly goods). Households in the East are significantly more likely to buy organic foods (by 12 per cent), to recycle paper (by 19 per cent) and to have cut down on car use (by 20 per cent).

Is there a typology of green behaviour?

The simple technique of identifying responses which are significantly higher than the sample average is again used to judge whether a typology of green behaviour can be identified. Those characteristics most associated with green behaviour, ranked by the number of 'significant' responses to the nine options shown in Table 3.4, are:

1. University education	4
2. Further Education	3

No other variable scored significantly above average on more than two questions. This implies that it is as difficult to identify respondents who consistently 'act green' from knowledge of their background as it was to identify those with consistently green beliefs. These findings are of note for what they do not reveal about the pattern of environmental behaviour in Scotland. Based on a regression analysis of behaviour in Britain, Witherspoon (1995) found other variables, in addition the two measures of educational attainment identified above, were positively related with green consumerism: general environmental concern, a willingness to sacrifice for the environment, concern for animal rights, support for the Liberal Democrats or Green Party and being a woman rather than a man. While there is no parallel for most of these variables in the 1998 Scottish survey, differences between the sexes did *not* emerge as particularly significant. That there is no gender gap to speak of may be surprising, given the split in household decision-making around shopping, and encouraging.

Conclusions

Environmental concern in Scotland reaches more widely than activism. The more specific any proposal to improve the environment, and the more clearly it is costed, the lower the level of public support appears to be. There is also evidence of inconsistent and contradictory attitudes, suggesting that green beliefs are wider than they are deep-rooted. One task for Scotland's new Parliamentarians will be to put down stronger roots, by raising awareness about the costs of inaction as well as the trade-offs involved with various policy options.

If more households are to be convinced of the need for change, responsibility for the costs must be clearly located. Where externalities can be traced back to the polluter, the costs should be allocated accordingly. Where car use or non-renewable energy consumption are targeted, the benefits of reform should be articulated. As David Begg argues in his discussion of transport policy in Chapter Six, the phasing of reform matters. Before consumers are asked to pay more in charges or taxes, some of the benefits should be brought forward. The best way to convince a sceptical public is to demonstrate that new money will be spent on the priorities it was raised for. If lasting progress is to be made, policy-makers will have to overcome their resistance to ear-marking revenue streams for particular expenditures.

The election of a government in 1997 which appears committed to tackling the growth of car use and switching resources from road-building into public transport improves the prospects for environmental sustainability. But it offers no guarantees. Politicians in Scotland and the rest of Britain will only compete to deliver better policies in the long-run when they believe there is a political 'market-place' for the environment.

In the meantime, institutional changes are needed to make the sustainable choices easier - meaning more accessible and affordable. Raising awareness about recycling is of dubious worth if only a small minority of households have regular collections from their homes, while two-car households ironically make extra short-distance journeys in order to recycle. Among the steps that might encourage more households to 'do the green thing' are:

- Regular collections on the doorstep would broaden the base of those able to participate in recycling. Recent examples of local authorities cutting their commitment to recycling should therefore be reviewed. The Prime Minister has called for every local authority to adopt a Local Agenda 21 strategy by the start of 2000. Access to the basics like recycling should be consistently improved.

- Local authorities should have a responsibility to send details about recycling and saving energy to every household.
- A modest Council Tax rebate could be paid on a trial basis to reflect the frequency of recycling by each household.
- Supermarkets and other companies (including the utilities) could adapt their affinity cards to allow shoppers to earn higher bonus points for buying green products.
- The cost of organic foods could be reduced through more effective organisation of the market. In particular the Scottish Office's Organic Aid scheme could be extended to help more farmers diversify production and encourage community food projects to establish trading links with high street supermarkets.
- The Scottish Parliament could set a target of increasing the volume of organic sales and the proportion of goods grown locally.

Securing a step change in awareness and activity will require the commitment of more than the Parliament. There will be no simple solutions for Holyrood to apply. Sustainability makes demands of citizens and consumers, employers and producers, as well as politicians in each tier of government. Devolution can make a difference if it adopts a problem-solving focus and designs a better mix of incentives than currently exists.

Notes

1 In March 1998, a series of questions on environmental beliefs and behaviour was included in the Scottish Population Omnibus conducted by Market Research Scotland on a monthly basis. The survey was based on household interviews with a sample of 1013 adults drawn from across Scotland. Further details are available from the authors.

References

Harrington, A. (1997) 'Scots green but not so cabbage looking', *The Herald*, November.

McCaig, E. and Henderson, C. (1995) *Sustainable Development: What it means to the General Public*, The Scottish Office, Edinburgh.

Wilkinson, D. and Waterton, J. (1991) *Public Attitudes to the Environment in Scotland*, The Scottish Office, Edinburgh.

Witherspoon, S. (1995) 'The greening of Britain: romance and rationality', *British Social Attitudes - The 11th Report*, Dartmouth, Aldershot, 107-139.

Witherspoon, S. and Martin, J. (1993) 'What do we mean by green?', *British Social Attitudes - The 9th report*, Dartmouth, Aldershot.

4 The Scottish Greens in a Changing Political Climate

LYNN BENNIE

Introduction

The Scottish Green Party has lacked political influence since its formation two decades ago. While the impact of a Green Party is difficult to measure, compared to many of its European counterparts, the Scottish party has been particularly marginalised and powerless, and to a greater extent than its sister-party south of the border. However, many believe that the establishment of Scotland's Parliament will herald the beginning of a period of multi-party politics which will provide new opportunities for small parties. This chapter details the history of the Scottish Greens, outlines those factors which explain the party's lack of success, and assesses the implications of devolution for the party. Are the Greens in Scotland entering a new period of political significance?

The Greens in Scotland

The Green Party in Scotland developed rather more slowly than the English party. The origins of Green Party activity in England can be dated back to 1973 with the creation of 'People', but Scottish Greens did not begin to mobilise until the end of the 1970s. The first Green Party election candidates in Scotland stood in 1979 in Edinburgh, five years after Greens contested local elections in England (Rüdig and Lowe 1986). Activists involved in the formation of the Scottish party felt that the environmental movement needed a strictly political wing to ensure that green issues were discussed during elections. At that time they had no clear expectations about how well they would perform in elections. They simply wanted the opportunity to challenge the positions of the other 'grey' parties in Scotland and to heighten the electorate's awareness of green issues. During the 1980s, branch activities were concentrated in the cities of Edinburgh and Glasgow and in the far North. Membership climbed very slowly, from 100 members in 1980 to around 300 in 1985, while electoral returns stayed stubbornly low (with the party polling 1-2 per cent of votes cast).

The end of the 1980s brought a period of relative electoral success for the Scottish Green Party. The growth of environmental consciousness across Britain led to a rise in all types of green behaviour, from recycling to green voting (Rüdig et al 1996). The high-point of electoral success for the Scottish Greens came in the 1989 European Elections. Green candidates stood in all eight constituencies in Scotland and received 115,000 votes, 7.2 per cent of the total Scottish vote. This was more than they had ever achieved in a national election, but not enough to win representation under Britain's first-past-the-post electoral system, and less than half the share of the vote won by the Greens in the rest of Britain (14.9 per cent). The three years between 1989 and 1991 was a period of relative success for the Scottish Greens. Membership of the party in Scotland rose to a high of 1,250, branch organisations reached a peak of 36, a Scottish Green regional councillor was elected in Highland, and the party gained its formal independence from the UK party.

In the 1992 General Election the Scottish Greens fielded 23 candidates and received just over 3 per cent of the vote in those constituencies they contested. The issue of Scotland's constitutional future dominated the election, and some felt that the party caused itself political embarrassment by using the UK Green Party's political broadcast. The party experienced a corresponding decline in membership levels and activity rates. From 1,250 members in 1990, membership fell to 225 within two years, most party branches ceased activity, and the party's one regional councillor defected to the SNP. The 1992 General Election had a devastating effect on Scottish Green activists, not only because the party itself had such a poor result. The fact that Labour lost meant that hopes of a Scottish Parliament were dashed.

During and following the 1992 election the Greens participated in broad alliances (Common Cause, Democracy for Scotland, and Scotland United) to generate grassroots support for constitutional change through public demonstrations, vigils and petitions, which contrasted with the elite-led Scottish Constitutional Convention's drafting of a constitutional package for change. The Greens left the Convention in 1990 because of a failure to agree on the principles of a multi-option referendum and a timetable for moving towards a system of proportional representation, but returned to the negotiations in 1993.

The Scottish Greens labelled 1994 'Year Zero', based on the hope that the party had reached its lowest point and could go on to recover. In the 1994 European Elections it was able to offer a full slate of candidates in all eight constituencies, but only due to financial assistance from the UK Greens. The party attracted 1.5 per cent of the vote, a disappointing result having achieved over 7 per cent of the vote in 1989. The Greens suffered a collapse in their European Election vote but the performance of the Scottish Greens was poor relative to the rest of Britain. The Greens won 3.5 per cent in England. No wonder the party was demoralised. At a very poorly attended 1994 Annual Conference, a demoralised Green Party debated the idea of withdrawing from the next General Election. It was suggested that the Greens in Scotland faced a hopeless fight under the current electoral system and that to fight and lose badly again would actually harm the development of the party by leading to another haemorrhaging of members. Other Greens suggested that refusing to fight the next General Election could be seen as a matter of principle, a protest against the electoral system and lack of devolved power in Scotland.

In the 1995 elections to the new unitary councils, the Scottish Green Party put forward 22 candidates, winning an average 4.8 per cent of the vote in contested wards. One Green candidate polled more than 10 per cent of the vote, but this was in a ward with only two candidates (North Fort William and Inverlochy). In the 1995 Parliamentary by-election in Perth and Kinross, the Scottish Greens won 5 per cent of the votes, but finished behind the Monster Raving Loony Party. Meanwhile the UK Greens won 22 local election seats in that year.

In 1997 Scottish Green participation in the General Election was minimal. The party put forward five candidates, failing to attract more than 1.5 per cent of the vote, and averaging 0.8 per cent. Despite the significance of the election result, Scotland's constitutional status was not as big an issue as in 1992. The election was characterised by voter fatigue and political sleaze. The environment was not a prominent issue and the Greens were even less conspicuous than in 1992. In the devolution referendum campaign, the party was fully supportive of the creation of a Scottish Parliament, but once again it had a very low profile. The referendum campaign was characterised by a lack of local campaigning and the dominance of the national media (Jones 1997, p.14), factors which undermined the role of small parties like the Greens.

It is clear that the Greens have been almost completely marginalised in recent historically significant events in Scotland. Currently, the party only has around three hundred members and it fails to register in public opinion polls. Four local parties remain active - Highland, Dundee, Edinburgh and Strathclyde. While the UK Greens have experienced a genuine taste of electoral success at the local level, the Scottish Greens have not. The party would argue that it is unwise to measure success by focusing on electoral performance and that its existence is justified by the radical nature of its agenda. Activists believe that the party is the only organisation in Scotland promoting a truly radical ecological philosophy, based on decentralised green economics which challenges the nature of the current political system while pressure groups only aim to reform it.

Unfortunately for the Green Party in Scotland, environmental pressure groups dominate green politics north of the border. The Scottish environmental movement is very diverse in its objectives, levels of influence and tactics employed. Its growth has resulted in a plethora of groups competing for resources, membership, media attention and access to policy makers, giving them a much higher profile than the party. Although there are examples of community led campaigns in which the Scottish Greens have been involved, it is the environmental pressure groups that attract media and public attention. There have been many recent examples of grassroots environmental direct action, much of it involving Earth First!, originally formed in the USA. This group has been actively protesting in Scotland since the beginning of 1994 and joined in protests against road-building throughout Britain, including Twyford Down and the M11. These protests have included attempts to stop the M77 Glasgow to Ayr road, disruption of super-quarrying in the Highlands, protests against Hunterston nuclear power station, and at the headquarters of Scottish Nuclear in East Kilbride. The non violent direct action (NVDA) employed by Earth First! has received consistent media attention.

One of the most widely publicised protests in recent years has been against the building of a new motorway through the South West of Glasgow, known as the Ayr Road Route or the M77 extension. A number of green groups were involved in the campaign. Glasgow Earth First (GEF), the Scottish Wildlife Trust, Friends of the Earth Scotland, the Railway Development Society, Glasgow for People (formed specifically to campaign against the construction of motorways in the city), World Wide Fund for Nature, and the Scottish Green Party all publicly supported the campaign, primarily with back-up support such as the delivery of food parcels and public statements of support. While the media took much interest in the M77 case, there was very little coverage of the Green Party's response to developments, preferring instead to highlight the radical tactics employed by Earth First!. These grass-roots campaigners have shown a willingness to confront authority at various levels, sometimes pushing the concept of Non-Violent Direct Action (NVDA) to its limit. The M77 activists were involved in mass demonstrations, they disrupted Regional Council meetings and they physically obstructed the felling of trees. Many of the protesters were arrested and charged by the police. Similar tactics have been used by groups protesting against nuclear power plants and a super quarry in the Highlands and Islands. The Green Party may be reluctant to associate itself with the more radical activities of the other sections of the environmental movement, for fear of discouraging potential voters, but this also makes it less 'media-genic'.

Whether it concerns road-building, the future of Scotland's water industry, or the construction of a funicular railway in Aviemore, it is environmental groups like Earth First, FoE and WWF that appear to be campaigning most effectively. Environmental groups dominate media coverage, while the Green Party appears rather peripheral. Over-shadowed by environmental pressure groups, and with no form of electoral representation in Scottish politics, the Scottish Green Party has failed to keep up with the UK Greens who had 26 councillors in principal authorities following the 1998 local elections.

Explaining the Development of Green Politics in Scotland

If support for Scotland's Green Party was measured as a strength of environmentalism, one would be forced to conclude that green politics in Scotland is essentially irrelevant. However such a conclusion could not be further from the truth. Environmental groups continue to promote environmental issues quite successfully. But why, exactly, has the Green Party in Scotland performed so poorly? The answer lies in the cool climate of Scotland's political system which offers few opportunities for the Greens. Rootes (1995) identifies three main factors which have been used to explain the development of green parties in Europe. These are the public's environmental consciousness, the opportunities and constraints imposed by institutional arrangements and the balance of political competition (Rootes 1995: 232). These factors are relevant to the Scottish case.

The rise of British environmental consciousness during the 1980s has been well documented (Bennie and Rüdig 1993; Rüdig et al 1993; Rüdig et al 1996). This has resulted in an increase in general awareness of issues like the greenhouse effect and the threat to the ozone layer, although by the time of the 1992 General Election environmental issues had disappeared almost completely from the political agenda (Carter 1992). In Scotland attitudes appear to have followed a similar path. A 1991 study of attitudes commissioned by the Scottish Office Environment Department provides evidence that these issues had emerged as important (Wilkinson and Waterton 1991). According to this survey, the Scottish public were most concerned about the pollution of rivers, lochs and seas and raw sewage put into the sea, followed by the quality of drinking water, nuclear waste and damage to the ozone layer. A comparison of Scottish attitudes with those in England and Wales in 1989 revealed broad similarities, the top two issues of concern being identical, but Scottish respondents indicated that they were more likely to 'act green' by using ozone friendly aerosols, picking up litter and so on (Wilkinson and Waterton 1991: 69). In Scotland, as in the rest of the UK, by the time of the 1992 General Election these concerns had been overtaken by traditional economic issues. The economic recession changed the political agenda, with debate focusing on more traditional concerns like unemployment and interest rates, and the passionate debate on Scotland's future helped ensure that environmental matters were pushed to the sidelines. In 1997 the environment appeared no more salient as an electoral priority.

However, the environment's lack of prominence at election times does not necessarily mean that it is unimportant. Although more research into Scottish public attitudes is needed there is no reason to assume that concern for environmental issues is particularly low in Scotland, as compared with the rest of the UK. Fairly extensive media reports of environmental incidents keep the environment in the minds of Scottish people. These have included the recent controversy over the importing of nuclear waste from Georgia to Dounreay, disposal of oil rigs at sea, the Braer disaster off the coast of the Shetland Isles, a recurring debate over the safety of Scotland's nuclear plants, and the reporting of international developments such as the French Government's nuclear tests in the South Pacific and the International Conference on Climate Change in Kyoto. The poor performance of the Scottish Green Party cannot be explained away by a low level of environmental consciousness in Scotland.

A more likely explanation for the disappointing performance of the Scottish Greens is the first-past-the-post electoral system which discriminates against small parties with low levels of support, especially when it is spatially dispersed rather than concentrated. Under this system the Greens have found it impossible to achieve electoral representation at the national level (even with over 7 per cent of the vote in 1989) and at the local level they have faired only marginally better. The Greens themselves argue that they will perform better in elections based on a system of proportional representation, although it is unclear to what extent the party will benefit under the Additional Member System (AMS) to be used in electing the new Parliament. The majoritarian system of British elections has certainly not helped small parties like the Scottish Greens. Nor do they receive public funding. In national and European elections, parties in Britain have been entitled to free distribution of a campaign address and radio and television broadcast time on the basis of strength in the outgoing Parliament and the number of constituencies being contested. Other than this, there is no public funding of political parties. Each General Election candidate is required to pay a £500 deposit, a large investment for a small number of votes. For a party like the Scottish Greens, reliant on its members for funds and unlikely to retain election deposits, the cost of contesting elections can be debilitating.

The lack of media interest is another significant problem for the Greens. North and south of the border, the Green Party has been ignored by the media. Only in 1989, during the European Elections, was the party given wide-spread media exposure. Since that time, a number of other small parties have competed for the attention of the media, including the Referendum Party and the UK Independence Party. Greens accuse the media of a 'green embargo' which 'impedes open debate and undermines political progress by restricting its attention to the main parties' (Barnett 1997, p.7). The lack of media coverage of the Greens is incontestable, but the media is still prepared to report on environmental issues like the decommissioning of oil structures in the North Sea and anti-road protests.

These factors help explain why the party tends to be left out in the cold without a great deal of support from environmental groups. The Scottish Greens have been constrained by a rather inhospitable institutional structure, as have the UK Greens. However, the Greens in Scotland face some unique difficulties. Scotland's party system has been a powerful constraint on the Greens. Because of the strength of the Scottish National Party in Scotland (SNP), the Greens struggle to make an impact on what is effectively a four party system. The SNP wins up to one third of votes cast in elections, leaving little room for a small party like the Greens to make an impression. In 1997, only 1.4 per cent of the total Scottish vote went to 'other' parties. The UK Greens do not have to contend with the extra 'nationalist dimension', and this fact alone probably explains why they have consistently done better in elections.

Another important feature of the party system in Scotland is the apparent 'greening' of the political parties. All the major parties in Britain have become wise to the popularity of the environment (Robinson 1992). This is also evident in Scotland. The Scottish Labour Party's 1997 election manifesto included commitments to return Scotland's water services to local democratic control, a review of Scotland's roads programme, an integrated transport policy, and a new Parliamentary audit committee to monitor environmental progress across government departments. Labour's review of the Scottish roads programme includes a working group (together with Highland Council) to examine tolls on the Skye bridge. In office, the Labour Party has implemented a number of its green proposals, including the setting up of the Environmental Audit Committee, and the March 1998 Budget announced a gradation of vehicle excise duty to promote low-emission vehicles. However, the party has been criticised for not acting quickly enough on its promises to review road-building and develop an integrated transport system. The SNP advocates returning Scottish assets of Railtrack into public ownership and are committed to an 'integrated National Environment Plan' to achieve sustainable economic development. The SNP is ambiguous in its attitude to road building, noting the importance of 'improvements to the road network' in its 1997 manifesto. The party has a well publicised policy of unilateral nuclear disarmament, and would end nuclear reprocessing at Dounreay. Under an SNP government, the Skye toll bridge would be taken into public ownership. Even the Conservatives have been forced to play the green card by stressing that they are the party of conservation, although compared with the other main parties in Scotland the 1997 Conservative manifesto lacked detail on environmental policy. Rather bland commitments included the promise to 'protect the rural environment by continuing to pursue policies which enable rural communities and the rural environment to flourish simultaneously' (Scottish Conservative and Unionist Party 1997: 24).

Of all the major parties in Scotland, the Liberal Democrats appear most environmentally aware. According to a long-time Scottish Green Party member, they have produced 'the most comprehensive and coherent transport policy by a mainstream political party north of the border.....absorbing many of the principles of sustainable transport policy' (Spaven 1996, p.19). They advocate planning in order to reduce the need for transport, the use of subsidies to be paid by the Scottish Parliament to Railtrack in an attempt to switch freight from road to rail, and they discuss the possibility of road-pricing, although they would abolish tolls on the Skye Bridge. The 1997 Scottish Liberal Democrat General Election manifesto included commitments to introduce a 'carbon tax' on fossil fuels, cutting the annual car tax from £145 to £10 for cars up to 1600cc, and the introduction of a new Scottish Department of Natural Resources with responsibility for protecting the environment in Scotland. The position of the Liberal Democrats in Scotland has been strengthened by recent election results despite falling levels of support. With ten MPs representing Scottish seats at Westminster they can now claim to be the opposition in Scotland at least on this measure. It is clear that the Greens in Scotland function within a very competitive opportunity structure. Scotland's party system might prove to be *the* biggest obstacle for the Greens. Even as institutional factors become more favourable, the Scottish Greens will still face an uncertain future.

Until now, the analysis has focused on factors which the Greens themselves have very little control over. However, a full explanation of Green fortunes in Scotland requires more than an analysis of external factors. We must also look to internal factors. Although the Scottish Green Party has not experienced the same tensions as their English counterparts over the question of party structure and leadership, it has not been without internal difficulties. For a period (1990-1993) the party's internal organisation was nothing short of shambolic. Mainly due to chronic under-funding the party was unable to cope with the increase in membership in 1989/1990; the recording of membership broke down, and members simply drifted away. In response, the party reorganised its internal structure around the eight European constituencies and it now has a national register of members.

Scotland's New Democracy

Labour's sweeping General Election victory in May 1997 was won with less than half of the votes (43.2 per cent across the UK). In Scotland, Labour's 45.6 per cent of the vote secured 77.8 per cent of the seats. The result of elections to the Scottish Parliament will be more proportional. In May 1999, Members of the Scottish Parliament (MSPs) will be elected using an Additional Member System of voting. There will be 129 MSPs, 73 directly elected by first past the post, and a further 56 additional members to be elected proportionally on regional party or group lists. The regions - eight in total - will be based on current European constituencies. The introduction of the new voting system provides hope for the smaller parties who should gain representation if they win a minimum proportion of the vote in each constituency. While there is disagreement over which parties will benefit in the long term from the new system (see Curtice 1996 and Dyer 1996), it is unlikely that any single party will win a majority on its own and that Scotland will therefore see the development of a new style of coalition politics.

From its inception, the Scottish Green Party supported the creation of a Scottish Parliament as the first step towards independence. There have been times when the campaign for a democratic Scotland appeared largely responsible for motivating Green activists. It was fully committed to

Scotland Forward, the umbrella campaign for a 'Yes, Yes' vote in the 1997 referendum, and the result is said to have created a new set of opportunities for the Greens in Scotland. At the Scottish Green Party conference in June 1997, Convenor of the party Robin Harper referred to the creation of the new Parliament as a 'superb opening to help create a completely different, novel, open, responsive representative government that would make Scotland the envy of the world'. Equally, David Spaven (1996, p.20) has argued that:

> A Scottish Parliament elected by proportional representation would open up the prospect of smaller parties securing election. The Scottish Green Party is currently very marginalised, but could nevertheless play a significant role in a Scottish Parliament in circumstances where the balance of power rested with smaller parties.

The key point for the Greens is that voters will have two votes, and it is possible that while they vote for their traditional party in the constituencies, they may vote for other parties, including the Greens, when it comes to the regional list. The 'wasted vote' argument is diminished on the second vote and those people with some sympathy for the Greens could support them without abandoning their traditional party attachments.

The Greens in Scotland cite the case of Ireland, claiming that the party in Scotland will also benefit from a system of PR. From very small beginnings and without any major electoral breakthroughs (compared to the British 1989 boom) the Green Party of the Republic of Ireland (Comhaontas Glas) has grown steadily in numbers and political significance. Under the Single Transferable Vote (STV) electoral system, which in theory minimises wasted votes, Greens have been represented in the Irish Parliament for the last ten years and they have a number of county councillors and urban district councillors. In the 1994 European Elections two Green MEPs were elected in Ireland, and in the General Election of June 1997, with less than 3 per cent of the vote, the Green party doubled its representation in the Dail, winning two seats in Dublin. There was also speculation about the Greens holding the balance of power. Scottish Green Party representatives have argued that the electoral system is all that separates the two parties. It has been claimed that in terms of public support the Irish Greens have little more first preference support than the Scottish Greens but the STV electoral system, which gives voters the chance to rank candidates, gives the Irish Greens the advantage of being many people's second preference (McCabe 1994, p.1).

However, comparisons with the Irish case provide false hope for the Greens in Scotland. Ireland has 41 constituencies and 166 seats, all elected through STV, giving small parties more opportunities because in principle no votes are wasted. The Greens in Scotland are overly optimistic to claim that they are the party of second preference for many voters. In the Scottish Election Study of 1992 respondents were asked to name their second preference party: only 2 per cent of Scottish respondents named the Greens. Nevertheless, when asked directly about their feelings for the Green Party, 21 per cent of respondents said they were in favour of the party and 28 per cent said they were against it (with 44 per cent neither in favour nor against). It might be argued that a significant number of Scottish voters could be persuaded to support the Greens if it was no longer perceived as a wasted vote. Even so, the Greens in Scotland have had enormous difficulties attracting electoral support and there are a number of reasons why the new system is unlikely to result in any kind of break-through:

1. First-past-the-post will still account for a clear majority of seats in the Scottish Parliament (73 of the 129 MSPs). These seats are likely to be decided on the basis of traditional party loyalties.

2. For smaller parties to benefit, there needs to be ticket-splitting, or cross-voting, when voters support a different party with their second vote. The Greens hope that under the new system voters will be more likely to give a smaller party their second vote. However, there is little evidence to date that Scottish voters would choose the Greens. It is also possible that for some voters the first (constituency) vote may be a tactical vote and that their second vote is cast for their preferred party.

3. The general low level of support for the Greens means that there is no guarantee the party would make an impact under the AMS voting system. While a lower share of the vote is required to achieve representation through the regional list part of the system, there will still be an effective threshold which parties will have to surpass to ensure election.

4. There will be competition from other small parties. For example, the Scottish Socialist Alliance (SSA), a group of left-wing parties including Scottish Militant Labour (SML) and the Communist Party of Scotland (CPS), may make an impact in the Glasgow area. It contested 22 Parliamentary seats in 1997 and Tommy Sheridan of SML won 11.1 per cent of the vote in Glasgow Pollok, finishing ahead of the Conservatives and

Liberal Democrats. It is also possible that new parties will emerge. The establishment of the independent Highland Alliance in the summer of 1998 may lead to candidates standing for election to Holyrood.

5. In order to keep other parties out, alliances may be formed. For instance, in order to suppress the threat of the SNP, it may be in Labour's interest to strike a deal with the Liberal Democrats. In these cases, the second vote will be guided by the first, and other parties will find it difficult to make an impression.

6. The Greens may need to work with others. The party has acknowledged that it will need a strategy to deal with competition from other small parties who will be attracted to the new 'democratic spaces' under PR, and it has discussed the need for arrangements to avoid targeting the same constituencies. The Greens in Wales showed the way through an alliance with Plaid Cymru which resulted in them gaining a Liberal Democrat-held Parliamentary seat. The most likely alliance at the moment appears to be between the Greens and the parties of the far left. Whether red and green can work together has been a subject of passionate debate, following a proposal of electoral cooperation from the Scottish Socialist Alliance (SSA). However, fundamental tensions exist between the Greens and the far left and there would inevitably be problems working together. Moreover, the Greens in Scotland have had some problems working with other parties in the past: they temporarily withdrew from the Scottish Constitutional Convention in 1990 leaving the party without a voice in the debate over Scotland's constitutional future. Nevertheless, in support of devolution, the Greens have cooperated with other groups in different forums, such as the Scottish Civic Assembly, the Coalition for Scottish Democracy and the Vigil for a Scottish Parliament. They have also worked with other groups to promote certain policy goals at the UK level. For example the Road Traffic Reduction Bill was jointly promoted by Friends of the Earth, the Green Party and Plaid Cymru.

7. Fighting the elections will have financial consequences for the Greens. There will be three different sets of elections in Scotland in 1999 - local elections, elections to the Scottish Parliament and European elections - which will take place within a few weeks of each other. Elections require money for campaigning and candidate deposits. The party will focus on the European and Scottish elections because they will both be contested under party list electoral systems. For the European elections the party will need to

submit one list of candidates, costing £5000 (Scotland will be regarded as one electoral region), although they may have a chance of retaining that deposit as the threshold is to be lowered to 2.5 per cent of the vote. The Scottish Parliament elections are likely to cost £500 per seat in the first-past-the-post section, and £1,000 for each of the lists. Whatever the result of these elections, the financial cost to the party will be considerable.

Conclusion

Partly because the Green Party in Scotland has performed poorly in electoral terms, it appears that, relative to other European states, the 'new politics' cleavage has had little impact in Scotland. Studies of the Scottish electorate suggest that voting behaviour is based on a mix of materialism, class and national identity (Bennie et al 1997). It is also uncertain if the Greens in Scotland will benefit from Scottish devolution and reform of the voting system. The new electoral system does not greatly advantage a small party like the Greens. Nor does Scotland's highly competitive party system. The promotion of green policies by all the major parties in Scotland means that as institutional factors become more favourable, the party's overall 'political opportunity structure' may not be much more advantageous.

Until now voting patterns have not been a good indicator of underlying levels of concern for the environment. Environmental pressure groups remain active and popular in Scotland. The most environmentally aware individual may not wish to 'waste' a vote by voting Green under first-past-the-post, but may be persuaded under a more proportional electoral system. The outcome of the Scottish elections is largely unpredictable. Britain's electorate has already shown a willingness to vote for different parties in elections to different levels of government, as the 1989 European election result proved. The proposed system is completely new and the reaction of the voters is untested. Party attachment has changed over the last two decades and may do so again. If Scotland experiences coalition politics, the Greens would not look as out of place as they currently do. Institutional change in the form of a new Parliament and electoral system may indeed create opportunities for the Scottish Greens, but they may have to be prepared to cooperate with other groups and parties, learning from the 'pragmatic realism' of the Greens in Wales. At the very least, the Scottish Parliament, and the opening up of the policy-making process, provides another potential route for getting environmental issues on the political agenda.

Notes

1 Since September 1990, there have been two Green parties operating in the UK: the Scottish Green Party and the Green Party of England, Wales and Northern Ireland. For reasons of simplicity, in this chapter I refer to the latter as the UK Greens.
2 The seats contested were: Caithness, Sutherland & Easter Ross; Edinburgh Central; Edinburgh Pentlands; Inverness East, Nairn & Lochaber; and Ross, Skye and Inverness West. The UK Greens put forward 95 candidates, returning an average of 1.4 per cent in these seats.
3 Along with the activities of the Scottish Socialist Alliance (SSA).
4 It is very unlikely however that the party will maintain this position following the elections to the Scottish Parliament under the Additional Member System.
5 These are Central Scotland, Glasgow, Highlands and Islands, Lothians, Mid Scotland and Fife, North East Scotland, South of Scotland, and West of Scotland.

References

Bennie, L.G. and Rüdig, W. (1993) 'Youth and the Environment', *Youth and Policy*, No.42.

Bennie, L.G., Brand, J. and Mitchell, J. (1997) *How Scotland Votes: Scottish Parties and Elections*, Manchester University Press, Manchester.

Barnett, P. 'Gagging the Greens', *Green World*, Summer 1997, No.18.

Carter, N. (1992) 'Whatever Happened to the Environment? The British General Election of 1992', *Environmental Politics*, Vol.1 No.3.

Curtice, J. (1996), 'Why the Additional Member System has won out in Scotland', *Representation*, Vol.33, No.4.

Dyer, M. (1996) 'Scotland's Additional Members and the Maintenance of Labour Power', *Representation*, Vol.34, No.2.

Jones, P. (1997) 'A Start to a New Song: The 1997 Referendum Campaign', *Scottish Affairs*, No.21, Autumn 1997.

McCabe, D. (1994) *Scottish Green Print*, June.

Mitchell, J. and Bennie, L.G. (1997) 'A Very British Affair? The 1997 General Election in Scotland', paper presented to Elections, Public Opinion and Parties (EPOP) Conference, September 1997.

Robinson, M. (1992) *The Greening of British Party Politics*, Manchester University Press, Manchester.

Rootes, C (1995) 'Britain: Greens in a cold climate', in Richardson, D. and Rootes, C. (1995) *The Green Challenge: The Development of Green parties in Europe*, Routledge, London.

Rüdig, W. and Lowe, P. (1986) 'The Withered 'Greening' of British Politics', *Political Studies*, Vol.34, pp.262-284.

Rüdig, W., Franklin, M.N. and Bennie, L.G. (1993) Green Blues: The Rise and Decline of the British Green Party', *Strathclyde Papers on Government and Politics*, No.95.
Rüdig, W., Franklin, M.N. and Bennie, L.G. (1996) 'Up and Down with the Greens: Ecology and Politics in Britain, 1989-1992', *Electoral Studies*, Vol.15, No.1, pp.1-20.
Scottish Conservative and Unionist Party (1997) *Fighting for Scotland*, Scottish Conservative and Unionist Party, Edinburgh.
Spaven, D. (1996) 'Transport Policy and a Scottish Parliament', *Scottish Affairs*, No.17, Autumn.
Wilkinson, D. and Waterton, J. (1991) 'Public Attitudes to the Environment in Scotland', *Central Research Unit Paper*, Scottish Office, Edinburgh.

5 Sustainable Development in Scotland: Responses from the Grassroots

ELEANOR McDOWELL and DOUGLAS CHALMERS

Introduction

In terms of societal, political and economic transitions this century, few forces have changed human values and affected local, national and global agendas like the environment. Recent decades have witnessed the growth of environmental groups and a drive for sustainability. Following on from the UN Conference on Environment and Development in 1992 (better-known as the Rio Summit), sustainable development has become a core policy commitment. UNCED's Agenda 21, which set out a strategy for sustainable development, has been taken up at both national and local government levels. Agenda 21 was endorsed largely because bottom up solutions were viewed to have a greater chance of success than approaches imposed from the top down. But how effectively are local commitments translated into effective projects which last? What examples are there of good environmental practice at a community level in Scotland, and what barriers, if any, obstruct the course of community involvement?

This chapter focuses on local environmental initiatives in Scotland. The concept of local sustainability is explored through three case studies of grassroot initiatives based on in-depth interviews with local activists. The first case study considers the campaign to oppose plans to build Scotland's only toxic waste incinerator in Renfrew, Clydeside Against Pollution (CAP). This is followed by an evaluation of a grassroots campaign in Easthall, Easterhouse, in Glasgow. After years of struggling against the worst effects of damp housing, local residents set up a dampness task force which gained support from local and European governments to finance a unique solar housing energy conservation project. The final case study deals with the formation of the *Different Dundee* magazine. This project developed from a grassroots concern with the future of the city, and traces the growth of a sustainable ethos in response to a series of events within the local community.

The case studies are not selected as representative of grassroots campaigns in Scotland. Instead, they offer a glimpse of local pressures for a better quality of life initiated by ordinary people. These examples are important because they demonstrate the impact of local activism and the possibilities for change at a community level. However, it is important to note that local sustainability cannot be studied in isolation. It must be cast in wider social, economic and political terms. Formal democratic mechanisms have to be established to encourage participation and greater local control. The problems of adjustment from an essentially bureaucratic and centralised political model to one which is decentralised and more egalitarian are considerable.

Overall, the analysis of grassroots activity in Scotland is under-researched, although a study of participation in local decision making by Reid (1996) makes a useful contribution to this debate. Reid developed a series of case studies of communities aiming for positive change in their localities. A range of issues are explored including what residents considered to be the most important indicators of sustainable development (sometimes expressed as 'quality of life'). People were able to state their local needs and express their concerns on issues such as development plans and the future of their communities. Methods to encourage participation ranged from community information days, environment days, planning for real events, round tables and workshops. Common features associated with the success of the projects appeared to be the concept of ownership and initiative. Involvement is felt to be meaningful if it is based on the participants' own terms, or as equal partners in the process. Research into local authorities and their commitment to Agenda 21 is becoming more common (Young 1996; Freeman et al 1996; Tuxworth and Thomas 1996; Brand 1996). The role of local government should be emphasised. Recent studies assessing their response to Agenda 21 are explored at a later point in this chapter. Before moving on to the specific case studies, the concept of local sustainability is considered.

Local Sustainability

There has always been a human dimension to environmental issues and problems. From time immemorial humanity has had an impact on the natural environment (Bramwell 1989; Fairbrother 1970; Evans 1997). Equally, as earth's citizens we are affected in numerous ways as a result of bad environmental management, short term gain and environmental exploitation. Of major interest is why certain individuals develop an environmental consciousness. Some make decisions to behave in a more environmentally responsible manner (recycling, green consumerism) while others actively challenge decisions immediately affecting their lifestyles or communities. Many local conflicts have to confront extremely powerful opponents such as developers, government departments and transnational corporations.

Selman (1996) suggests that the acronym NIMBY (not in my back yard) has often been used in a pejorative sense to dismiss local opponents of apparently desirable schemes, by making them appear 'ignorant and selfish' when in reality they may be 'a legitimate response to externally generated demands on local assets' (p.3). Within this context it is vital to remember that principle 10 of the Earth Charter (included in Agenda 21) states that, 'environmental issues are best handled with the participation of all concerned citizens, at the relevant level'.

A key ingredient in local action must be the individual citizen. This presupposes that citizens will be motivated to work collectively on environmental principles and are actively encouraged to do so. Sustainable development is about empowerment, implying a role for everyone in an effort to bring social, economic and environmental benefits. Thus, in the wake of Rio there are some genuine calls for an increase in co-operation and participatory democracy among local communities. Turning to the first case study, the vigour of community action becomes apparent with the Clydeside Against Pollution Campaign.

Responses from the Grassroots: Three Community Case Studies

Clydeside against pollution

Scottish media attention has recently focused on the long-running Renfrew Incinerator Campaign. In July 1990 a public notice in the Scottish press confirmed fears that a toxic waste incinerator was planned for the town. (Mellor, 1992). This proposal by the Caird Group PLC to install a hazardous waste incinerator in Renfrew resulted in one of the biggest community based campaigns in Scotland. More than 10,000 'Clydeside Against Pollution' posters were displayed in house and shop windows and mass rallies and marches united the people of the town. The campaign won all party support and raised the issue of toxic incineration plants being sited in the heart of population belts, with the most senior environmental groups in Europe. After an eight year battle, the Caird PLC Group decided to withdraw their proposal in February 1997, insisting their decision was a 'commercial' one. The local campaigners on the other hand believe that their efforts were rewarded and considered the withdrawal a victory for the local community.

Environmental concepts based on a 'quality of life' approach have gained ground in recent years. In the Renfrew case, local people felt compelled to challenge the proposed developments. Concern over health risks[1], environmental pollution, and knock on effects to the local community were high on the agenda. There was also a sense of community protection, to assert control over the community's environmental history of industrial development and the subsequent pollution. Foster Evans, a leading activist and resident of Renfrew, spoke of the 'toxic sacrifice zone' where dirty industries are actually enticed. Evans questioned why people who have lived for decades with heavy industry should continue to be a target for unattractive industry.[2] The plans for the first commercial toxic waste incinerator in Scotland to be located in the heart of a heavily built up area close to homes, schools and businesses (such as food processing plants) were vehemently opposed. The possible health implications were of paramount concern to campaigners. Problems of asthma were already evident in the community and the threat of pollution reinforced this concern. Some may point to the employment opportunities the plant could have brought to Renfrew. However, with such overwhelming hostility to the development, if the community reached such a clear decision for its own benefit - and at its own cost - then these views should be respected.

A number of important factors advanced the campaign's cause. The efforts of a dedicated and committed group of individuals was fundamental. A petition was organised in September 1989 and Renfrew Against Pollution was formed (RAP). By October 1990 (RAP) was changed to Clydeside Against Pollution (CAP) to reflect the widening circle of opposition including the other side of the River Clyde, particularly Yorker. (Mellor 1992). In every sense, this was a bottom up campaign which galvanised a community into action - women, children, young people, pensioners. Important links were formed the business community, farmers, doctors, academics, church groups, community groups, tenants groups and local MPs. Over 7,000 people marched through the streets during a mass rally in October 1991.

Leafleting teams were organised throughout Renfrew, public meetings were arranged by local people, and signed petitions were sent to the Secretary of State for Scotland and the European Parliament. The campaign also gained extensive media support, with thousands of people attending CAP's Festival of the Environment held in Robertson Park (Renfrew) in

1993. There was a growing awareness over the transportation of toxic waste at sea and the wider impact of environmental issues, but the rise of the campaign in the late 1980s preceded the debate and implementation of Local Agenda 21.

Co-ordinating such a large campaign and maintaining the momentum over a number of years was problematic. Activists faced demands on their time from other campaigns but took the decision to remain focused on the immediate objective in Renfrew rather than risk being diverted by other causes. Advice and information was passed to other communities and CAP gained valuable support from campaigners as far afield as America, Londonderry and Doncaster.[3]

Various fund-raising methods were used and additional financial and practical support came from local businesses, local environmental groups, and community groups. Some funds still exist and plans are being discussed to commemorate the campaign with a plaque in Renfrew and Yorker, and the possibility of an exhibition to show the various stages of the campaign and the local involvement.

There are a number of similarities between the CAP campaign in Renfrew and the second case study based in Easterhouse. Both were grassroot initiatives where local residents challenged circumstances or decisions they perceived as having a negative impact on their quality of life. Both campaigns forged partnerships with other communities, agencies and professional bodies. The Easthall campaign was also a long and hard-fought battle but, as the case study outlines, it resulted in a remarkable achievement for the local community and in turn gained national and international recognition.

The Easthall solar housing initiative

In 1993 the Easthall community of Glasgow witnessed the completion of the first ever tenant-led Passive Solar Energy Demonstration Project (at a cost of £1.3 million). Greater Easterhouse is Glasgow's largest post-war public housing scheme. This extensive scheme was erected, in haste, on the north-eastern periphery of the city as part of a massive post-war slum clearance and re-building programme. Due to a shortage of resources, cheap and inappropriate industrial building techniques were used for the housing. In some neighbourhoods, little was provided by way of amenities for the large number of residents.

Other policy developments aimed at improving living conditions, such as the Clean Air Act (1956), led to more expensive forms of gas and electric heating (replacing open fires) which proved too expensive for poorer households. Subsequently, problems of cold, damp houses, poor transport facilities, a lack of other services, high unemployment, poor health (both physical and mental) resulted in serious hardship for many residents. A 1991 survey of housing conditions revealed that problems of mould, dampness and condensation affected 36 per cent of the housing units in Glasgow. (Brooke 1994). Recent evidence shows that:

> One third of households in Scotland (650,000-800,000) are unable to heat their homes for less than 10 per cent of net income and are living in or are at risk of fuel poverty (Revie, 1998, p18).

It is worth remembering that one of the Agenda 21 principles states that:

> Access to safe and healthy shelter is essential to a person's physical, psychological, social and economic well being and should be a fundamental part of national and international action.

Glasgow has the highest level of multiple deprivation than any other city in the UK. A third of the people in Glasgow are in receipt of social security benefits and 86 per cent of the most deprived postcode sectors in Scotland are located in the Greater Glasgow Health Board area. (Strathclyde Poverty Alliance, 1994). Numerous medical and sociological studies (Martin et al, 1987; Whitehead 1988, 1994; Carstairs and Morris 1991; Long et al, 1996) have established close links between deprivation and health. Poorer people tend to have higher morbidity and mortality statistics than the more affluent. Without disputing the importance of behavioural factors on health, there are nonetheless many structural constraints contributing directly to poor health, such as poverty, poor housing, occupational hazards and unemployment, which are largely outside the direct control of individuals. The Glasgow Healthy City Project (1995) stressed that 'health is not primarily a natural phenomenon. It is socially and economically produced and is, therefore, amenable to change (Matthews, 1997, p8).

The effect of bad housing on health was recognised by the Easthall Residents Association (ERA) which had been in existence for about twenty years. Although the Association dealt with a variety of issues affecting the community, the issue to dominate the agenda increasingly became damp housing. In an attempt to deal with the problem, a dampness task force was set up by the local residents. Throughout the 1980s and 1990s the health effects of bad housing was an issue which brought together similar groups throughout Scotland. Tenants lifestyles were often considered the cause of the dampness: they were unjustly accused of not sufficiently heating their homes. In response the tenants sought the help of professionals and other agencies for support and advice.

The communities take action Within Easthall the residents began to collect and swap information with other groups concerned primarily with health issues. These groups, largely from Glasgow and Edinburgh undertook their own research into the effect of housing on health, they interviewed councillors and were encouraged by community development workers to present their case to a research seminar at Edinburgh University. The liaison between local residents and academics resulted in a large scale study into the relationship between damp housing and health in three cities: Glasgow, Edinburgh and London. The studies showed statistically significant relationships between the problem of damp in houses and instances of respiratory, gastrointestinal problems, childhood infections, and emotional distress of residents living in these conditions after controlling for factors such as smoking and household income (Martin et al, 1987; Platt et al, 1989).

The Heatfest competition A local resident of Easthall, Cathy McCormack, who participated in the three cities housing research, was instrumental in organising a three day ideas competition called 'Heatfest' in January 1987. The aim of the competition was to devise ideas on how to improve the residents housing conditions. The ERA were members of the Technical Services Agency (TSA) a community technical aid centre which previously carried out in-depth surveys of the Easthall flats. The TSA had many academic links and, together with the ERA, launched Heatfest. The event attracted people from all over the UK: architects, surveyors, engineers, housing managers, university professors, and representatives of tenant groups from the West of Scotland. A crucial component of the Heatfest weekend was that for the first time, the people who live in damp houses got the opportunity to voice their views to a group of professionals on the on the kind of housing they would like to live in. Helene Martin, resident and chairperson of Heatfest summed this up:

> Hey look! .... we are not daft, we know how we want to live. We only need an opportunity like this to get our ideas and solutions on the drawing board, in a way that enables these people to understand that we have our own form of expertise (cited in McCormack, 1994, p. 151).

In 1989 practical ideas stemming from the Heatfest competition (which refined the best features from all the entries) were brought together in a successful bid for funding from the Commission of European Communities (provided moral and financial support was given from Glasgow District Council). The ERA campaigned relentlessly to obtain the necessary financial backing for the project, they organised a range of fundraising activities and tried to enlist the help of the private sector, the media, the Prime Minister, the Royal Family, the Scottish Office. Finally, a community conference was held to engage the support of active tenant groups who lobbied elected councillors all over Glasgow. Just before the EU deadline, Glasgow District Council consented to back the project, the Scottish Office gave them an additional borrowing consent of £750,000, and £370,000 came from their existing financial resources (McCormack 1994, p.152).

The Solar Housing Project in Easthall, Easterhouse, Glasgow It is notable that McCormack[4] stressed the 'genuine partnerships' which resulted from the solar housing initiative. A Management Committee consisting of ERA, Glasgow District Council, contract architects and the TSA was formed to run the project which was finally completed in 1992. With a range of energy conservation methods and solar energy technology, thirty six flats were super-insulated. Central heating and conservatory extensions were installed and solar panels on the roofs provided additional warmth and hot water. The verandas were glazed in, providing extra insulation and functioning as a reservoir of heated air for ventilation in the bedrooms. The result was warm, dry homes which were significantly cheaper to heat. Many residents claim their health has improved, and money saved on fuel bills can go towards other essentials such as an improved diet. Maintenance costs and condensation claims were also reduced (McCormack, 1994a).

The Heatfest solution had other wider environmental outcomes. In the first place, by using a clean, energy efficient source, the re-furbished flats were more than contributing their share of reducing the UK target of 25 per cent reduction of in $C0_2$ emissions - fifteen years ahead of schedule. Second, common links between impoverished communities in Glasgow and those of Third World countries were recognised. McCormack compared the effects of those in her own community who suffer 'the same pain and the same negative imagery as their third world counterparts' (1994, p.147). In 1996 she was sponsored by a number of organisations to go on a Health Study Tour of Nicaragua. Grassroot movements (Movimientos Communales) which included local children were considered an essential part of the local community.

Whether in Easterhouse or Nicaragua, grassroot responses mean taking action into your own hands. Much can be learned by sharing experiences between communities. Social and personal needs must be addressed as much as environmental needs. They are not mutually exclusive. When asked the question 'How would you define sustainability?', McCormack quickly responded: 'I can tell you what it isn't - it isn't wasted lives or wasted resources'.[5] She reflected on her previous housing situation, which caused her continuous stress and depression:

> I had to continually throw out mouldy furniture, clothes and toys and found myself having to choose between feeding my hungry children or hungry fuel meters which kept demanding more and more money (McCormack, 1994, p.148).

The Easthall solar housing project is highly significant as it represents a clear example of the stimulus for a major initiative coming from disadvantaged people themselves; and because the creation of partnerships with academics, local authorities and official agencies was a 'bottom-up' procedure driven by local campaigners passionate about the needs of their families and their communities. The solar housing project gained a great deal of attention locally, nationally and internationally. Researchers, academics, medics, environmentalists and planners have written to local residents or travelled to Scotland to view the solar houses. In 1994, Easthall activist Cathy McCormack was invited to address the UN Commission on Sustainable Development in New York as an official delegate for the Scottish Environmental Forum. At the Commission she stressed the potential of people at the grassroots who can bring tremendous benefits to their communities, if they are really listened to and heard.

The Easthall activists were successful in their campaign objectives. However, due to financial constraints, many other residents in housing need (including those involved with the campaign) were not beneficiaries of the solar housing project. Sustainable development depends on initiatives like that in Easthall with the genuine potential to improve health, cut costs and use energy reserves more efficiently. It is the scale of the problem which means urgent action spread more widely is needed. The true test is whether the lessons from Easthall will be applied more consistently.

At this point, the focus moves from Easthall to Dundee. Many of the factors pertinent to the outcome of the previous two case studies and those reviewed by Reid (1996) focused on individuals with a commitment to their localities. Advancing participation from the grassroots allowed residents to shape the process and assert some form of ownership, rather than having to react to a pre-set agenda. This scenario is also applicable to the third case study based on the *Different Dundee* initiative.

Different Dundee: Responses to Agenda 21

The *Different Dundee* magazine developed from grassroots concern for the future of the city. Now in its second year (and its sixth issue) with a quarterly circulation of 600, the magazine seeks to play an independent role in bringing Dundee City Council and local citizens closer to the concept of Local Agenda 21. The initiative for the magazine stemmed from a conference held in June 1995 on the theme *Different Dundee*. The conference was organised by a local political campaigning group and attended by the city's MPs, council leaders and local authority representatives. At this point, Local Agenda 21 did not particularly feature in the minds of the organisers, although they were aware of its existence. In conjunction with the broad discussions over the city's future, concern was growing over emissions from the local municipal waste incinerator at Baldovie. Opened 16 years previously, there was frequent opposition to the incinerator which burned Dundee's municipal waste together with some low level waste from local hospitals.

This opposition was focused through two events. First, in 1994 the European Community published figures outlining acceptable levels of emissions from waste incinerators. Second, the Environmental Protection Act of 1990 made it mandatory for owners of waste plants to collect and publish data on emissions. Tests at the Baldovie incinerator in 1993 revealed that the incinerator failed to meet the statutory standards. Researchers from the *Different Dundee* magazine spent time poring over 3,000 pages of tabulated data to highlight the suspected harm being done.

The results, published in an early issue of the magazine, caused a furore among the local community. Residents were alarmed that if breaching the new figures was deemed to cause harm, the cumulative harm in the past must be much greater, when higher levels were accepted. Calls for greater transparency grew along with a demand for an independent health survey (owing to concern about the levels of asthma and bronchial complaints in the locality). The *Different Dundee* magazine allowed local people to voice their opinion and simplified some of the technical material produced by Council departments.

The magazine's editorial board published a leaflet outlining facts regarding emissions (presenting the relevant information in two pages where the local authority had taken 600) and a community meeting was organised and attended by over 150 people. Comments from a local resident reflect the atmosphere of the meeting:

> The council officials present were genuinely amazed that a local community could be so well informed about an issue which until then had been shrouded in rather technical jargon.[6]

The agenda shifted towards the control of local residents to which the Council had to respond.

In isolating reasons why the campaign took off, it is notable that the core of the campaigners were also involved in other community issues. A recent victory had been won in preventing a local public house with a reputation for rowdiness getting its licence renewed. Thus an existing network of community activists had just tasted success when *Different Dundee* published the figures on EC emissions.

Due to the achievements of *Different Dundee*, the Council sponsored the magazine to organise a 'Dundee - Our Future' conference to consider aspects of sustainability. The conference was a success and the Council passed a motion supporting the Agenda 21 process and the formation of a Sustainability Forum. The Forum produced a professional sustainability action plan in April 1997 and contributed to raising the awareness of Agenda 21 within the community. Another conference on Sustainable Development in the Built Environment is planned for late 1998.

By December 1996, the Baldovie incinerator was closed. The Council failed to upgrade the technology required to meet the new regulations, opting for a more modern incinerator. An investigation into the effects of the original incinerator was also established, covering the impact on residents' health. In the plan of decentralisation the new Sustainability Forum is officially recognised as a 'Community of Interest' together with neighbourhood forums which the Council has encouraged. Issue four of *Different Dundee* marked the completion of the magazine's first year in existence. Reasons for the magazine continuing into its second year were considered in the editorial pages:

> There is need for a thoughtful commentary on where the city is going. 'DD' can be a locally based space for people to think out loud and assist in providing allies for those they want to make changes to the effects, and nature of society as it is (*Different Dundee*, 4. p.3).

Thus in Dundee, through the efforts of a local magazine and the active involvement of the local community, the door remains open for a continuing and more effective bottom-up participatory process in keeping with the overall spirit of Agenda 21.

Grassroots Activity: The Way Forward?

There are various examples of community activity in Scotland evidenced by the three case studies. Recognition and sharing of good practice will hopefully facilitate progress in other Scottish communities. Progress, however, is unlikely to happen in isolation. Normally there is a need for co-operation and communication between grassroot groups and other 'community of interest' groups, the media, professionals, academics and local government, termed by some theorists as 'alliances for sustainability' (Hutchcroft, 1996). Solutions can be found among local communities and others can learn from their experiences in an effort to encourage debate, share information and experiences, raise awareness, gather funds, and galvanise support. Information, however, must be consistent, open, accessible and comprehensible.[7]

Commitment, patience and enthusiasm are prerequisites for campaigners, along with the aim of working towards a collective purpose and objective. Another important variable is the prevailing public opinion on the issue. In the case of Clydeside Against Pollution, there was overwhelming support from the media, political parties, local business and neighbouring communities. Immense challenges nevertheless confront grassroot groups. In some instances they have to confront very strong opposition. There is also the social acceptability of the issue - will the media and general public be sympathetic? Other problems include a lack of funding, a lack of political will and excessive demands on time which can demoralise even the most committed of campaigners. Although similarities can be drawn between different grassroot initiatives, they are by their very nature unique in terms of: the key protagonists, their tactics, the prevailing issue at hand, the level of support, the timing and the final outcome. There is no standardised approach to sustainability, but there is a growing recognition of the need to challenge traditional ideas and practises.

Chapter 23 of Agenda 21 states: 'One of the fundamental prerequisites for the achievement of sustainable development is broad public participation in decision-making' (FoE, 1996, p.79). Although the concept of empowerment is theoretically gaining ground (Golding 1994; Young, 1996) there are doubts surrounding the reality of this in practise. Voisey and O'Riordan (1997) suggest that:

> To involve citizens effectively requires changes in the patterns of power, representation, and a transparency of operations that political parties are not willing to condone at present (p48).

Munton (1997) echoes the view that central government, despite its rhetoric, remains unsympathetic to the logic of empowerment contained within Agenda 21.

A system which truly valued community empowerment could pave the way for a more sustainable future. As the case studies suggest, inclusive and flexible measures must be found which suit the respective needs of communities. The greater the community involvement, the greater the erosion of the 'democratic deficit' in the pursuit of development that is economically efficient, socially just and environmentally sound. Further decentralisation and a fuller recognition of the importance of participation, quality of life and equity are increasingly being voiced (Scandret, 1997).

Recent demands have also been made for the rights of children to be included in issues affecting their quality of life with the development of youth councils and forums (Matthews and Limb, 1998). Others advocate the participation of children in environmental planning, suggesting that children would offer a fresh perspective and, through their active involvement, strengthen their sense of environmental responsibility (Adams and Ingham, 1998). It is worth reinforcing that McCormack (in the second case study) drew attention to the grassroots Movimiento Communal in Nicaragua which traditionally includes children. Other types of participatory measures include young people's parliaments, citizens' juries, consensus conferencing and mediation groups (Matthews, 1996). Participation must cut across the boundaries of class and race, health and illness, gender and age, as well as geography. Nitin Desai, (Under Secretary General of Policy Co-ordination and Sustainable Development, UN) stressed that:

> ....It is not possible to have a policy framework which attacks one dimension, by delivering the problem of basic needs without addressing the other dimensions of marginalisation from the political process, of social discrimination or even rootlessness (Matthews, 1997, p.8).

When it comes to analysing the distributional consequences, both social and spatial, lower income groups are more vulnerable to environmental problems, unwanted development and a lack of meaningful consultation in the overall process of decision-making (Lowe, 1977).

There might be more room for optimism if opportunities for greater community participation are maximised by the Scottish Parliament. A recent report by the John Wheatley Centre suggests that the new Parliament will:

> Work to ensure that the public are given a full opportunity to take part in environmental decision making and securing public trust over the condition of our environment (John Wheatley Centre, 1997, p11).

Sustainable development must be high on the agenda for the new millennium. The Scottish environment has many distinctive attributes which generate specific Scottish requirements. The Holyrood Parliament will have the opportunity to deal with issues 'closer to home' and fine tune decisions and policy proposals more relevant to Scotland. With this in mind, it is essential to connect the joint goals of environmental and social improvement. Dunion (1995) underlines this point suggesting that 'A Scottish Parliament should not become the vehicle for economic development which ignores equity and the environment (p39)'.

The Role of Local Authorities

The UN Conference on Environment and Development reported in 1992 that 'because so many of the problems and solutions being addressed by Agenda 21 have their roots in local activities, the participation and co-operation of local authorities will be a determining factor in fulfilling its objectives' (UNCED, 1992, Chapter 28.1).

In what sense can local authorities develop a comprehensive and co-ordinated response towards sustainable development in Scotland? The popular slogan 'think globally act locally' stresses the connection between personal and community action with larger scale environmental pursuits, but these slogans are not always backed up with the explicit mechanisms needed to put them into effect. The inter-connectedness of environmental problems requires conditions that enable co-operation and partnership to take place.

Christie (1996) claims that a major prerequisite of sustainable development is 'strong local government' and a commitment to 'decentralised environmental decision-making and local consensus building' (p.29). In addition, the balance of power which has been eroded in local authorities must be restored. Reference is made to the 'internal management upheavals and the inflexibility of cost-led compulsory competitive tendering' (Christie 1996, p31). The disruptive impact of the fragmentation of local authorities and the effect on the capacity to actively pursue sustainable development has generated much concern (Patterson and Theobald, 1995; Stoker 1996).

A convincing argument from Selman and Parker (1997) suggests that citizenship is about power and the local arena is a legitimate setting for the exercise of this power. As such, LA21s should harness the energies necessary to engage widespread participation in local sustainability and quality of life issues. Principle 10 of the Earth Charter, Agenda 21 states that:

> ....Mechanisms should be created to facilitate the active involvement and participation of all concerned, particularly communities and people at the local level, in decision making on land use and management, by no later than 1996.

Six years have passed since the Rio Earth Summit where 180 governments pledged to introduce sustainable development strategies. The section below considers British local government participation in the LA21 process.

Agenda 21: Recent studies of local authorities

The Local Government Management Board (LGMB) has conducted annual surveys of Local Agenda 21 activity in British local authorities. The results of the 1994/95 and 1996 surveys have been analysed by Tuxworth and Brown (1996). Findings of the latest report showed that 90 per cent of UK local authorities were committed to Local Agenda 21, with almost half pledging a strong endorsement. When asked whether sustainable development was viewed as a 'significant influence' in driving policy, responses were variable. Over 50 per cent viewed it as having a significant impact on land use planning, energy management and transport, 31 per cent on economic development and regeneration, 23 per cent on anti-poverty strategies, 22 per cent on housing and 14 per cent in relation to welfare and equal opportunities. Some authorities were using innovative methods of involving the local community: 25 per cent were using focus groups, 20 per cent adopted visioning or future search exercises, and 13 per cent used planning for real and parish mapping exercises.

A range of views were also sought from local authorities in terms of their 'level of impact' on specific areas related to sustainable development. Highest ranked by impact achieved were, 'resource use; limiting pollution; beauty/distinctiveness of the area, and biodiversity.' Conversely, 'equity' issues such as 'meeting human needs, and achieving satisfactory work for all' were viewed as having least impact. Given the importance of the 'quality of life' dimension associated with Agenda 21 and the impact of this variable on the Scottish case studies, this finding gives cause for concern.

Apart from local authorities, a broad spectrum of bodies actively seeks public and government participation in the pursuit of a more sustainable Scotland, including charitable/not-for-profit organisations; environmental pressure groups; research institutes; and private companies. The growth in such activity is a sign of health. However the value of this work will only be maximised if local government is *more consistent* in its commitment to LA21. The new Parliament is unlikely to be impressed by Scottish councils which are slow to form alliances, and are unwilling or unable to maintain them. It may set clearer targets for partnership working for sustainability, resulting in more and better local data. While non-statutory agencies will emerge and fade over time, local government will have an ongoing responsibility for civic leadership. It can frustrate the efforts of others but it also has the capacity to promote sustainable alliances.

Conclusion

Sustainable development has many dimensions, cutting across economic, social and cultural spheres, challenging traditional assumptions and the centralised decision-making of experts. There is a general lack of awareness over the actual concept of 'sustainable development' among the lay public (McCormick and McDowell, 1998). However, as the quality of life dimension in personal and collective identities continues to grow in importance, sustainable development incorporating moral and social concerns may have a greater appeal at a grassroots level. Clydeside Against Pollution, the Easthall Solar Housing Project and *Different Dundee* were all driven by powerful quality of life issues. Despite the relative deprivation in Easthall, the local community fought to secure a unique sustainable initiative. All of the case studies signified the broader potential of bottom-up strategies, and alerted us to the fact that communities which have no access to participation are inherently unsustainable.

Collaborative and inclusive practises must be encouraged to facilitate the development and achievement of local sustainable initiatives. We are increasingly forced to accustom ourselves to an environmental interdependence at local, national and international levels. As Selman (1996) ambitiously states: 'Issues of sustainability can encourage people to see themselves as 'global citizens' and to become more appreciative of their role within a web of international responsibilities' (p161). If the ultimate aim is not to compromise the environmental prospects for future generations we should willingly accept the challenge.

Notes

1 The possible exposure to dioxin (a widely-feared pollutant) was at the heart of opposition to the proposed Renfrew incinerator.
2 Interview between Foster Evans (leading CAP activist) and Eleanor McDowell, 10 June 1998
3 Ibid.
4 Interview between Cathy McCormack and Eleanor McDowell, 5 February 1997.
5 Ibid.
6 Interview between Editorial Board of *Different Dundee* and Douglas Chalmers, 1 February 1997.
7 Valuable information is given in *Protecting our Environment: A citizens guide to environmental rights and action in Scotland* (Friends of the Earth Scotland, 1993).

References

Adams, E. and Ingham, S. (1998) *Changing Places: Children's Participation in Environmental Planning*, The Children's Society, London.
Agyeman, J. and Evans, B. (1995) 'Sustainability and democracy: community participation in Local Agenda 21', *Local Government Policy Making*, 22 (2), pp. 35-40.
Aldous, K. (ed, 1996) *Who's Who in the Environment: Scotland*, The Environment Council, Edinburgh.
Bramwell, A. (1989) *Ecology in the 20th Century: A History*, Yale University Press, London.
Brand, J. (1996) 'Sustainable Development, the International, National and Local Context for Women', *Built Environment* Vol 22, No 1. (SLPU)
Brooke, J. (1994) 'Towards an Ecological City? Poverty and urban regeneration in Glasgow' in Bhatti, et al (eds) *Housing and the Environment: a new agenda*, Charted Institute of Housing, London.
Carstairs, V. and Morris, R. (1991) *Deprivation and Health in Scotland*, Aberdeen University Press, Aberdeen.
Christie, I. (1996) 'A Green Light for Local Power', *Demos Quarterly*, (9).
Different Dundee, (1996) Issue 4, Autumn.
Dunion, K. (1995) *Living in the Real World: The International Role for Scotland's Parliament*, SEAD publications, Edinburgh.
Evans, D. (1997) *A History of Nature Conservation*, Routledge,
Fairbrother, N. (1970) *New Lives: New Landscapes*, Architectural Press, London.

Freeman. C. et al. (1996) 'Local Government and Emerging Models of Participation in the Local Agenda 21 Process', *The Journal of Environmental Planning and Management* 39. 1, pp. 65-78.

Golding, A. (1994) 'Empowerment and Decentralisation' in J. Agyeman and Evans, B. (eds) *Local Environmental Policies and Strategies*, Longman Publications.

Friends of the Earth Scotland (1993) *Protecting our Environment: A citizens guide to environmental rights and action in Scotland*, FoE Scotland, Edinburgh.

Friends of the Earth Scotland (1996) *Towards a Sustainable Scotland: A Discussion Paper*, FoE Scotland: Edinburgh.

Hutchcroft, I. (1996) 'Local Authorities, Universities and Communities: Alliances for Sustainability', in *Local Environment*, 1, pp. 219-224.

Long, G. McDonald, S. and Scott, G. (1996) *Child and Family Poverty in Scotland: The Facts* (second edition), Save the Children/Glasgow Caledonian University.

Lowe, P.D. (1977) 'Amenity and Equity: A review of local environmental pressure groups in Britain', *Environment and Planning*, Vol. 8, p.35-58.

Matthews, H. and Limb, M. (1998) 'The right to say: the development of youth councils/forums within the UK', *Area*, 30.1, pp. 66-78.

Matthews, P. (ed) (1996) 'Greening the Grey: Setting the Urban Agenda for the 21st Century', *Proceedings of a Conference on Habitat 11 and Sustainable Settlements*, Scottish Environmental Forum.

Matthews, P. (1997) *Poverty and Sustainable Development in Scotland: Meeting the Challenge of Implementing Chapter 3 of Agenda 21*, Scottish Environmental Forum.

Martin, C. et al (1987) 'Housing conditions and ill health', *British Medical Journal*, 294. pp. 1125-1127.

Mellor, D. (1992) *Burning Issue*, Clydeside Against Pollution.

Munton, R. (1997) 'Engaging Sustainable Development: Some Observation on Progress in the UK', *Progress in Human Geography*, Vol 21, 2, pp. 147-163)

McCormack, C. (1994) 'From the Fourth to the Third World: A Common Vision of Health', *Conference Proceedings of the European Health Policy Conference: Opportunities for the Future*, WHO, Copenhagen, pp.146-154.

McCormack, C. (1994a) *On the Global Agenda: SEAD Annual Report*, pp. 8-9, Scottish Education and Action for Development, Edinburgh.

McCormick, J. and McDowell, E. (1998) 'Environmental Beliefs and Behaviour in Scotland', in McCormick, J. and McDowell, E. (eds.) *Environment Scotland: Prospects for Sustainability,* Ashgate, Aldershot.

O'Riordan, T and Voisey, H. (1997) 'Governing Institutions for Sustainable Development: The UK's National Level Approach' in O'Riordan, T. and Voisey, H. (eds) *Sustainable Development in Western Europe: Coming to Terms with Agenda 21*, pp 24-54.

Platt, S. et al (1989) 'Damp Housing, mould growth and symptomatic health state', *British Medical Journal*, 298. pp. 1673-1678.

Patterson, A. Theobald, K. (1995) 'Sustainable Development, Agenda 21 and the New Local Governance in Britain', *Regional Studies*, Vol 29.8 pp. 773-778.

Reid, D. (1995) *Participation in local decision making*, Scottish Natural Heritage, Perth.

Revie, C. (1998) *The Impact of Fuel Poverty and Housing Conditions on Scotland's Health*, Energy Action Scotland.

Scandret, E. (1997) 'Tools for a Sustainable Scotland', *Scotland 21 Today*, Issue 5, Nov/Dec, CSV Environmental Publications.

Selman, P. (1996) *Local Sustainability: Managing and Planning Ecologically Sound Places*, Chapman.

Selman, P and Parker, J. (1997) 'Citizenship, Civicness and Social Capital in Local Agenda 21', *Local Environment*, Vol 2, No 2, 1997.

Stoker, G. (1996) *Rethinking Local Democracy*, MacMillan, Basingstoke.

Strathclyde Poverty Alliance (1994) *Communities Against Poverty: Resource Pack*, SPA, Glasgow.

Tuxworth, B. and Thomas, E. (1996) 'Local Agenda 21 Survey', compiled for the Local Government Planning Board, LGMB, Luton.

Walker, P. (1995) *A Woman's World: A pull out guide to Agenda 21* and *Women Turning Dreams into Concrete Reality*, New Economics Foundation, London.

John Wheatley Centre (1997) *Working for Sustainability: An Environmental Agenda for a Scottish Parliament*, Report of the Commission on Environmental Policies for a Scottish Parliament, John Wheatley Centre, Edinburgh.

Whitehead, M. (1988) 'The Health Divide', in Townsend, P. Davidson, N. and Whitehead, M. (eds) *Inequalities in Health*, Penguin, London.

Whitehead, M. (1994) 'Health, Housing and Settlements', *Conference Proceedings of the European Health Policy Conference: Opportunities for the Future*, WHO, Copenhagen, pp. 155-168.

Young, S. (1996) 'Stepping stones to empowerment? Participation in the context of Local Agenda 21' *Local Government Policy Making* Vol 22. No 4, March.

6 Creating the Framework for Sustainable Local Transport Strategies in Scotland

DAVID BEGG

Introduction

The establishment of the Scottish Parliament in 1999 presents the opportunity for re-thinking how transport policy is delivered in Scotland. This chapter aims to stimulate discussion on how best to arrange the strategic transportation planning responsibilities, and to demonstrate the potential benefits that could accrue from the opportunities for Scotland to make its own decisions in this area. There can be little doubt that there is a pressing need for the planning and management of Scotland's transport system to be overhauled. Perhaps a better word is *established*, since no integrated strategic management platform has previously existed. In this context a Scottish Parliament should provide an opportunity for improvement rather than a catalyst for further confusion.

Responsibility for transport in Scotland is currently split both by transport mode and by geography. Opportunities for co-ordination have been limited and diminishing. Local government reorganisation and the imposition of legislation which encourages fragmentation and a go-it-alone mentality have had a devastating effect on the ability of planners to address transportation strategy and related land use development issues effectively.

The aims of this chapter are:

- to examine why the transport infrastructure in Scotland is increasingly unable to cope with the demands placed upon it;
- to explain why the opportunities for strategic planning have been diminished by local government re-organisation and government policy;
- to discuss the opportunities for transport planning arising from the Scotland Bill and to present a blueprint for transportation planning under a Scottish Parliament.
- to look at the issue of under-investment and alternative sources of funding for transport.

Problems with Current Policies

Increasing demands on a weakening infrastructure

The increasing dependence on motor vehicles, and especially the private car, is the biggest transport problem facing the UK today. Over the last 20 years the number of households in the UK with regular use of a car has increased by over 50 per cent. Even more alarming is the fact that almost a quarter of households now have access to two or more cars. Not only are there more cars on the roads. People are also travelling further: over the last 20 years, average journey distances to work, shopping and education have increased by between 35 per cent and 40 per cent. As a result the less well off, children and young adults, the elderly, and those who are unable or unwilling to drive are increasingly being denied opportunities.

What are the implications of increased car use in this country? Problems are worst in our cities, but evidence from England suggests that the major inter-urban routes are themselves becoming clogged. Scotland's problems are just a step behind our nearest neighbour. Some of the main issues to be addressed are:

- road safety: while accident casualty numbers are reducing, this is at least partly because people are restricting their own (or their children's) activities to avoid the risk of accidents;
- congestion: is now costing British industry £15 billion according to the CBI. It is not financially, physically or politically feasible to build new roads on a sufficient scale to cater for forecast traffic growth;
- pollution: road transport is the major contributor to poor air quality. It contributes 53 per cent of Nitrogen Oxide emissions and 90 per cent of Carbon Monoxide. Road transport is also a major source of noise pollution;
- finance: it is becoming increasingly difficult to fund either maintenance of the existing road network or investment in new roads.

Local government re-organisation

There is no doubt that local authorities in the urban areas now cover too small a geographical area and have insufficient resources to plan and implement transportation strategies effectively. To some extent this is recognised: responsibility for strategic planning through the Structure Plan process has been kept with larger groupings such as Lothian, and Glasgow/Clyde Valley. The retention of a Passenger Transport Executive covering the former Strathclyde Region and those covering the six English conurbations recognises the fact that big cities dominate travel patterns over an extensive hinterland and that passenger transport networks and services need to be planned and delivered on that scale.

The only coordination mechanism for urban areas outside Strathclyde is the Structure Plan process, which relies on a system of voluntary co-operation. This does not, however, give direct influence over the management of transport and, given that it has yet to be tested, its effectiveness remains open to question. The crunch will come when a difficult decision has to be taken over a major planning application or in the next review of the structure plan. Can there be consensus among unitary authorities flexing their muscles and ready to defend the powers given to them by the Local Government etc. (Scotland) Act 1994? There is a serious danger that divisions will quickly emerge between authorities and one or more may choose to go their own way and ignore the views of the rest. There can be no question that the arrangements inherited by the new Parliament will be far weaker in ensuring a cohesive region-wide approach to planning than those which existed under the previous regional councils.

Fragmentation has had financial implications as well. The new authorities surrounding Edinburgh and Glasgow have very limited budgets available for transport investment and support. Would it, for example, have been possible to build the new M8/M9 link road in West Lothian or electrify the North Berwick line in East Lothian at costs of £10m and £0.7m respectively from those local authorities' annual capital budgets of around £1m? The transport 'Challenge Fund', intended to provide funds for larger projects, will only support a limited number of schemes and requires a substantial resource outlay to develop a bid.

Scottish Parliament White Paper

Following the clear public endorsement of plans for a Scottish Parliament in the devolution referendum in September 1997, decisions on transport, economic development, land use planning, local government and the environment will be made in Edinburgh rather than Westminster from the year 2000. The Parliament will also decide how to allocate the Scottish expenditure block and will have the power to vary the basic rate of income tax set by the UK Parliament by up to 3p.

Legislative devolution presents a number of advantages for transport planning. Decentralising power from Westminster should increase the democratic accountability of decision-making, ensuring that policies target Scotland's needs. The Holyrood Parliament will have the opportunity to improve the way in which transport policy is delivered, including funding mechanisms, grant assessment procedures and the administrative framework. There are, however, two specific concerns about the Scottish Parliament's transport responsibilities which the White Paper raised.

The first relates to the Parliament's tax raising powers. By limiting its fiscal powers to varying the level of income tax, the UK Government treats taxation simply as a revenue raising tool, ignoring its role as a means of influencing behaviour. This runs contrary to the policy of recent administrations which have used the tax system to discourage consumption of cigarettes and alcohol, reduce the amount of waste (the landfill tax) and, by increasing fuel duty, encouraged the purchase of fuel efficient cars.

The Labour Party's 1997 UK general election manifesto (Labour Party, 1997) said that the use of the taxation system in this way would continue

> Taxation is not neutral in the way it raises revenue. How and what governments tax sends clear signals about the economic activities they believe should be encouraged and discouraged, and the values they wish to entrench in society. Just as, for example, work should be encouraged through the tax system, environmental pollution should be discouraged.

If the Scottish Parliament is to reflect Scottish needs then it should have the ability to vary taxes to reflect the different economic, social and geographical conditions north of the border. For instance, because of the higher than average petrol prices paid in rural Scotland, the Parliament might wish to increase fuel duty taxation at a lower rate than the 6 per cent per annum which the UK Government is committed to. According to Hart (1997):

> It is arguably more important that a Scottish Parliament should have the power to vary road fuel duties and supervise road pricing than power to vary income tax ... It is hard to see how effective, sustainable and flexible transport ... policies could be applied without such powers. They are the outstanding omission from the White Paper - admittedly reflecting a UK reluctance to get to grips with issues of transport transparency, environmental costing and road pricing.

In its taxation proposals, the White Paper revealed itself to be behind the times in thinking on economic and environmental issues.

A second area of concern was the White Paper's treatment of rail services. It appears, however, that these are to be remedied. Under the original proposals, the Parliament would have had some rail grant powers but overall financial support to rail services in Scotland will remain the responsibility of the Office of Passenger Rail Franchising (OPRAF) in

London. Given that the vast majority of Scottish rail services are operated by the Scotrail franchise, there are strong logistical, as well as democratic, arguments for transferring control of these to a Scottish Parliament. This was acknowledged by the then Devolution and Transport Minister Henry McLeish in Spring 1998, proposing that the Scottish Executive:

- Should be able to issue objectives, instructions and guidance in relation to passenger rail services which both start and end in Scotland. This will enable Scottish Ministers to instruct the Government's proposed new rail authority on levels of service, fares and how the ScotRail franchise should be managed;
- Will have responsibility for the funding of rail services in Scotland;
- Will be able to issue objectives, instructions and guidance in respect of ScotRail sleeper services, subject to this advice not impacting adversely on the rail authority's costs outside Scotland or the operation of rail service generally;
- Will be able to issue non-binding guidance in respect of other cross-border services;
- The Scottish Parliament should have legislative competence to determine the extent of the rail and bus responsibilities of Strathclyde PTA/E and of any new such bodies;
- The Scottish Parliament will have legislative competence over the powers for the promotion and construction of new railways in Scotland;
- The Government will bring forward new legislation in the context of their Integrated Transport White Papers.....to transfer these responsibilities to the Scottish Parliament and Executive. In addition, in terms of the Scotland Bill, the Parliament will have legislative competence over grants for passenger rail services and the Scottish Executive will have responsibilities for freight facilities grants and track access grants in Scotland;
- Scottish Ministers will have responsibility for appointing the Chairmen of the Rail Users' Consultative Committee for Scotland. The reports of the Committee and the Central Rail Users' Consultative Committee will be laid before the Scottish Parliament (Scottish Office, 1998).

The Government has also promised that the appropriate public expenditure transfer will be made by OPRAF to the Scottish Executive.

A Regional Framework for Scottish Transport

Transport partnerships for Scotland

A Scottish Parliament needs to have full powers in strategic land use planning, economic development and all forms of transport as well as a meaningful funding base. Such a framework would allow the development of an integrated, sustainable approach to transport in Scotland. However, it would be difficult (and inappropriate) for a Scottish Parliament to exercise these powers in a centralist manner. The Parliament based in Edinburgh should not itself be formulating strategic plans for Aberdeen and its hinterland or even for that matter in the Edinburgh area. On the other hand, as has already been outlined, the new councils, certainly in the urban areas, are too small geographically to implement such plans in isolation.

A new type of organisational and decision-making framework is therefore needed to manage and develop transport in Scotland in a sustainable manner. A strategy is needed to link the responsibilities of central government, a Scottish Parliament, local authorities, Passenger Transport Executives and the Local Enterprise Companies. Each of these bodies has a relevant input to the development of strategic plans for the major geographical units that make up the country.

A transport map of Scotland is now beginning to emerge consisting of five 'Transport Partnership areas'. The Strathclyde PTE area (perhaps embracing Dumfries & Galloway), South East of Scotland, North East Scotland, Tayside (including Dundee, Perth & Kinross and Angus) and Highland and Islands. The North East and South East transport priorities could be financed by a combination of urban road pricing and Private Non-Residential (PNR) parking taxes; the Strathclyde area or South West could utilise PNR taxation (local politicians have argued that road pricing would be difficult in Glasgow); and I would contend that Highland should be given a large proportion of any revenue that is made available from the fuel tax escalator with some top-up from PNR taxation in towns such as Inverness.

This five-area Transport Partnership model for Scotland should not just apply to local authorities but should also be a blueprint for a decentralised structure for the Scottish Parliament. There is no reason why local authorities and the Parliament cannot work closely together when it comes to delivering transport services. I would envisage local authority councillors and Members of the Scottish Parliament, as well as the appropriate officers, serving on the new Transport Partnerships. The Partnerships should have responsibility for rail and bus services, trunk roads and strategic planning within their boundaries. Such a structure will be crucial if the Government is to achieve its integrated transport objectives.

Funding Transport Investment

A transport fund

The introduction of a Scottish Parliament gives the opportunity for fundamentally changing the funding arrangements for transport to try and correct some of the difficulties outlined earlier. Cities throughout the world that have made major transport investments have usually done so through a mix of central government funding, private investment, and local taxes such as sales and payroll taxes. The current position in the UK where 85 per cent of local authority spending is typically controlled directly by central government militates strongly against imaginative or innovative action.

A 'transport fund' would pool existing sources of funding, private sector inputs, and the revenue that could be generated from transport-related charges and taxation, in particular new charges designed to balance the cost of transport to the user with the external costs imposed on the community at large in the form of pollution, congestion, healthcare and emergency services. Only by establishing such a transport fund can a rational investment appraisal be created, providing a level playing field between the different modes of transport and different types of area. This is an essential prerequisite for a coherent national transport policy. Some of the potential funding sources are now discussed.

Existing Funding Sources

The public sector

In total, public expenditure on transport in Scotland amounted to £806m in 1994/5. This excludes the rail Passenger Service Obligation (PSO), which can only be estimated from the national figure, but could be around £100m pro-rata population. Out of an estimated total of £900m spent on Scottish transport in 1994/95 therefore, £242m (27 per cent) was spent on the trunk road network, and £442m (49 per cent) on local authority roads. Capital expenditure accounted for £381m, virtually all of which being spent on roads. These figures do, however, exclude investment in rail improvements which are recouped through fares and revenue support to the rail industry.

The private sector

Private sector money can also contribute to the implementation of transport measures. However, there are severe limitations to the attractiveness of such investment to the private sector, as well as a view that such investment may only replace public capital expenditure with long-term repayments from public revenue budgets. Only projects with a direct revenue stream are attractive to the private sector, and trying to package such schemes in with non income-producing projects has proved unacceptable. This may distort investment towards projects that generate income rather than those which produce the highest return to the community as a whole in terms of environmental improvement, economic regeneration or other welfare criteria.

New Sources of Revenue

There are other ways in which public authorities could themselves raise additional revenue. By introducing charges either on employers, on road users themselves, or on local consumers, it could be possible for local authorities both to correct the perceived relative costs of alternative modes, and to provide income for investment in better transport. For new revenue-raising powers to be publicly acceptable, and to achieve the objective of a better transport system, there are two essential prerequisites:

- Acceptance of the principle of *hypothecation* to ensure the funds are used for transport-related purposes rather than other services.
- Acceptance of the principle of *additionality*, ensuring that the funds raised are not simply substituted for other sources of funding.

The taxes, levies and charges listed below are variously used in other countries for financing transport. Before any of these were seriously considered for implementation, there would clearly need to be a thorough examination of all the possible impacts and requirements, including:

- Likely effect on travel behaviour
- Consequences for all transport modes and for land use
- Impact on business, the property market and major services
- Assessment of boundary or cordon effects
- Impacts on different sections of the community
- Technological aspects
- Administrative and legal requirements

Even with the introduction of additional revenue-raising powers, it may be necessary in some of these cases to phase expenditure to ensure that at least some aspects of better public transport facilities are in place before introducing any financial arrangements that effectively restrain car traffic as well as generating income.

Charges on employers

Uniform Business Rate (UBR): With control of the UBR passing to the Scottish Parliament, a share of the revenues could be identified specifically as a 'transport rate'.
Payroll levy: A percentage levy on the payroll of all employers (perhaps only above a certain size). Equivalent of the 'Versement transport' levied in France.
Business subsidy of public transport season tickets: A variant of the above, providing a more direct link between the employer payment and a transport benefit.
Private non-residential parking levy: Private Non Residential parking provision free to the user is a distortion in the perceived cost of car use, and a not inconsiderable perk in city centre locations.

Vehicle user/owner charges

Vehicle licence supplement: A supplementary charge for use of a vehicle within a city. The boundary within which the licence is required could be either the city centre or a larger area.
Fuel tax: A percentage of fuel tax, or an urban fuel tax supplement, could be allocated to local authorities for transport purposes.
Road tolls: This could range from simple bridge/tunnel tolls, to outer or inner charging cordons, and eventually to full-scale road pricing. The first hint of this was found in the Government's White Paper on Integrated Transport for Scotland, suggesting tolls on the M8 Motorway between Glasgow and Edinburgh.

Sales taxes

Car sales levy: A levy on all car sales within a defined area.
Sales tax: A local retail sales tax, or possibly a percentage of VAT revenue on sales within the city.
Visitor tax: Levy on bed-nights at hotels, guest houses etc.

Conclusion

The Scottish Parliament offers opportunities for more effective transport planning in Scotland, with greater accountability and policies more suited to the nation's priorities. Though government policy omits powers to use the taxation system to influence behaviour, it creates the potential for real subsidiarity in which decisions affecting particular types of area are taken in the appropriate forum for that area. Thus matters affecting the whole of Scotland, such as the necessary framework for achieving sustainable development, should be taken by a Scottish Parliament. Those issues only affecting particular communities, such as local planning matters, should be taken by local councils. But matters affecting whole city-regions and Scotland's substantial rural regions should be taken by bodies representing these regions as a whole.

The missing piece in the jigsaw is the ability to plan the provision of transport strategies at a regional level. Rather than recreating the old Regional Councils as a further tier of government, a more appropriate response would be for the Parliament to adopt a decentralised approach to its transport powers. A series of Transport Partnerships could cover Scotland, involving representatives of the Parliament and local government. The Partnerships would also oversee strategic land use planning issues. These could be an innovative and effective method for giving life to the principles of subsidiarity and sustainability.

References

Hart, T. (1997) *Transport needs from a Scottish perspective - towards consensus and clarification.*

Labour Party (1997) *Britain deserves better*, General Election Manifesto, Labour Party, London.

Scottish Office (1998) 'Devolution settlement to bring major railways responsibilities to Scotland', News Release 0679/98, Scottish Office, Edinburgh.

7 Planning and the Parliament: Challenges and Opportunities

SARAH BOYACK

Introduction

In order to set the background to the possibilities offered by the establishment of the Scottish Parliament in the field of planning, it is worth discussing what sort of planning system we currently have and how well it is working, before going on to look at the specific opportunities and challenges that will be posed by devolution. A brief overview of the current operation of the Planning System in Scotland is provided, highlighting current problems and areas where there is room for improvement. Some of the opportunities which are then discussed may come through the current initiatives being implemented by the Labour Government, others from the creative policy melting pot that will be the new Parliament.

The notion of town planning in its current form is one of the survivors of the 1945 Labour Government's radical programme. The collective experience of wartime and the need to rebuild large areas of our towns and cities provided democratic legitimacy for an activity which encouraged long-term thinking and co-operation between those who had the ability to develop land for the benefit of the community. Over time the system has changed, proving its ability to reflect the priorities and ideology of the government of the day. However, the survival of town planning as an activity during the 1980's and 90's, while under sustained attack during the Thatcher years, is notable. From Michael Heseltine's tirade against planners who kept jobs locked up in their filing cabinets, to John Gummer discovering the positive contribution that planning could make in tackling the new challenges of sustainable development, a remarkable turnaround occurred. The balance swung from planning being seen as interference in developers' rights to make money, to the discovery of the benefits of a system with the potential to protect individual householders' rights. Most of these battles took place in England with less notable confrontation between communities and developers in Scotland. A major source of dispute in England was routes for new roads, in Newbury, Stonehenge and Birmingham for example. With the important exception of the M74/77 proposals for motorway extensions in Glasgow there has not been the same experiences of direct action against new roads proposals north of the border.

In Scotland, while there have been notable battles on specific sites such as the Harris Superquarry, there have been few high profile disputes about development. There is less sense of the planning system being stacked against individuals or developers. However, the quality press have been taking an increasing interest in planning issues in relation to green belts, flooding, playing fields and opencast mining. Communities have become more articulate and better organised in protesting against developments they feel are not in their interests.

The Current Planning System in Scotland

The current planning system in Scotland essentially leaves central government to provide advice and guidance to local authorities with the Secretary of State having specific powers of approval of structure plans and acting as a final arbiter in disputes over planning applications. The system is set out in the 1997 Consolidated Planning Act. It relies on local authorities to deliver planning services within the broad parameters set by central government. This enables local input and accountability as authorities employ planners to provide professional advice which is then considered by councillors when making decisions on planning matters. However individual planning decisions are not made on party political grounds as they are required to reflect the development plan policy framework and material considerations, i.e. relevant planning matters. The emphasis on the system being 'plan-led' was underlined by the 1991 Planning and Compensation Act which applied to England and Wales and is now reflected in the Town and Country Planning (Scotland) Act 1997. Section 25 of the Act states that:

> Where, in making any determination under the planning acts regard is to be had to the development plan, the determination shall be made in accordance with the plan unless material considerations indicate otherwise. [1]

The development plans system comprises *structure plans* which set the strategic framework for housing, transport and economic development, and *local plans* which translate the strategic framework into local policies and proposals and deal with site specific allocations for future development. The Scottish Office sets the framework for plan preparation through the series of National Planning Policy Guidelines (NPPGs), Planning Advice Notes (PANs) and Circulars. The Scottish Office approves structure plans, and has a system of reporters who preside over Local Plan Inquiries to give a hearing to those objecting to the content of local plans.

Local planning authorities are required to consider proposals to develop land or change its use. Applicants have the right to appeal although 92 per cent of all applications are approved and less than 2 per cent result in appeals, only one third of which are upheld.[2] The Secretary of State is the final arbiter on all decisions with appeals being considered by Scottish Office Reporters and the Secretary of State having the power to override their recommendations if he chooses. The Secretary of State can also call in applications and recall appeals where he feels there is a national planning interest. There are also designated areas where planning decisions are expedited in order to stimulate growth for specific local areas called *simplified planning zones.* In practice however their use has been extremely limited, with only four such areas designated in Scotland.

Effectiveness of the System

The Citizen's Charter introduced by the Major Government was extended to advice from the Scottish Office with expectations set out for developers, members of the public and government with terminology used in PAN 40 stressing the importance of 'customer care' and standards of service.[3] But there have been long-standing concerns expressed by government and the business community about the system's speed and effectiveness. These concerns prompted a consultation exercise by the Scottish Office in 1994 in the run-up to local government reorganisation (The Review of the Town and Country Planning System).[4] Although the broad conclusion of the Government's response was that the system worked, there were areas where improvements were required. Subsequently, the Scottish Office commissioned a series of research projects to address the key components of the system: Review of Planning Enforcement; Review of the General Permitted Development Order; Review of Development Planning in Scotland; and Evaluation of the Effectiveness of National Planning Guidance. In addition in 1997, the Scottish Office initiated a Development Control Audit of three local authorities on a pilot basis in order to explore different planning authority practices and ascertain the scope for speeding the system up. Advice in PAN 40 on Development Control identified good practice for adoption by local authorities.

The role of development plans in providing a framework for the system is acknowledged both by professionals and politicians alike as being crucial to the successful operation of the system. Structure plans are meant to look ahead to a ten year time horizon, with a review carried out after five years. Local plans are meant to have a five year shelf-life with regular review and monitoring taking place. However, in the *Review of Development Planning in Scotland* the evidence suggests that the timescales of development plan production and approval are too slow. Analysis of approval and preparation timescales showed that, for local plans, '42 per cent had been finalised more than five years previously, but with no finalised replacement by this date'. Furthermore a sample of Scottish local plans showed that the average preparation time from the beginning of the process to adoption was 5.5 years. While there is broad coverage of Scotland by local plans, they tend not to be up-to-date or regularly reviewed. Whereas structure plans have tended to be reviewed and monitored on a more regular basis, the Review identified problems at the final stage of approval where plans submitted to the Secretary of State took on average fourteen months to be approved.[5] Given that structure plans take around three years to be prepared this means that the data on which they are based are already becoming outdated by the time that the plans have been finalised. This has a knock on effect for local plans and by the time they have implemented structure plan requirements, too much time has elapsed.

The research team concluded that the credibility and effectiveness of the planning system was being undermined by the delays in the system. Their final conclusion was that the whole culture of plan making needs to be re-energised:

> Plan-making to date could be characterised as embarking on a journey without fully mapping out the intended route, the ultimate destination, or how long it will take. There is no little irony in the chronic failure to effectively plan the *preparation* of a plan (and keep on '*planning the plan' thereafter).* There needs to be a renewed commitment to the ongoing process of plan-making if the plan-led system is to be delivered. [6]

Concerns about speed have not disappeared with the election of the Labour Government. Indeed the themes of a more focused, effective and modernised planning system have underpinned Ministerial speeches, both at a UK and a Scottish level. In February 1998 Scottish Office Minister for Planning Calum MacDonald summarised the Government's short term agenda for planning as being 'to modernise it, to speed it up and provide an overall service that meets the needs of the user'. He also identified the need to link the planning system to the broader objectives of the Government:

> ...economic growth and social progress - that is Welfare to Work and tackling Social Exclusion.....a greater emphasis on a bottom-up approach

> through Community Planning...towards a better co-ordinated and corporate approach that not only seeks Best Value but promotes Added Value.[7]

Planning and the Parliament: the Potential Impact of Devolution

The White Paper on proposals to establish a Scottish Parliament provided few surprises in planning terms, as it closely followed the principles set out in the Constitutional Convention's scheme[8]. In effect the whole system transfers across to the Parliament as the system is already administratively devolved. Devolution opens up opportunities both to improve the current system and to think into the next millennium to decide what kind of Scotland is desired and what changes in the planning system are required to bring it about. Among the key issues are:

- progress towards a sustainable development policy framework;
- priorities;
- the delivery of strategic planning;
- political policy integration and delivery of transport, economic development and housing responsibility for final decision making;
- the role and involvement of the Scottish Office and civil servants;
- linkages between government funded agencies on implementation;
- the collection of data at a Scottish level which can be drawn on by local authorities;
- the need to discuss best practice throughout the UK; and
- European implications for Scottish planning.

Understandably, the White Paper said little about the operational arrangements of the Parliament, suggesting that this is a matter to be resolved by the Parliament. However there are important decisions to be made. Where will final accountability for decisions lie? How can the decision-making process be made more transparent? How will decisions be reached on planning applications of national importance? What will be the political mechanisms for discussing and approving planning legislation and policy guidance, decisions on development plans and planning appeals. In addition the role and responsibilities of quangos in relation to Holyrood and local government will need to be considered afresh.

The opportunity will clearly be there for more effective scrutiny of Scottish legislation and there will certainly be more time for introducing legislation compared with current arrangements at Westminster. At the moment it is difficult to secure space for legislation on Scottish matters in a crowded parliamentary agenda. This may have particular benefits in relation to planning matters which are rarely at the top of the political agenda unless there is a controversial decision to be taken.

Integrated policy making and delivery is opened up by devolution. The Scottish Environment Protection Agency (SEPA), Scottish Homes, Scottish Natural Heritage (SNH) and Scottish Enterprise are key players in the planning system. They could be required to meet the broad framework of the Parliament's policies on sustainable development. Implementation mechanisms also need to be considered. While accepting the warning from Professor Cliff Hague that 'the first debates should not be about administrative structures or policy mechanisms'[9], the structures and mechanisms are nonetheless important. If they do not match the purpose of the system, it will be much harder to implement strategic planning.

In the absence of a regional tier of government there is the issue of ensuring effective strategic frameworks for the whole of Scotland. There are several areas in the central belt where structure plans are prepared jointly by a combination of authorities. These new structure plan areas were established by the Scottish Office and split up the structure plan areas operated by the former Regional Councils in Strathclyde, Central, and Grampian. Different practice is developing and new arrangements are in place in order to ensure the strategic level does not disappear. In Ayrshire, the three local authorities have adopted a co-operative approach. South Ayrshire, East Ayrshire and North Ayrshire have established a Joint Structure Plan Team funded on an equal basis and are now working together to produce a plan for Ayrshire following the publication of an Issues Paper in 1997. In Glasgow and Clyde Valley, eight authorities have come together to produce a structure plan with a jointly-funded full time team and back up from individual authorities and joint officer teams. In the Lothians, however, joint working is carried out by individual members of staff across the four councils in the former Region (Edinburgh City, West Lothian, East Lothian and Midlothian), working together only on a limited basis. Several councils are responsible for a structure plan for their area apparently without the need for joint working. Argyll, Falkirk and Moray are examples.

Across Scotland there are now fewer staff resources available for plan preparation, implementation and monitoring. Local authorities are prioritising preparation of local plans to implement the requirements set out in the most recently approved round of structure plans. Current indications are that planning authorities are likely to prepare briefer, less comprehensive plans. The Scottish Office could assist in preparing a national digest of key data which would provide a national base line for structure planning review.[10]

While the capacity to deliver strategic planning has reduced on the local authority side, the necessity of liaison between different agencies with strategic responsibilities has increased. Local authorities no longer have within their own budgets the resources to deliver key aspects of their plans. This means that the issue of plan ownership is important and other stakeholders need to be more fully involved in the process of identifying planning priorities and strategies. The Local Enterprise Companies have an important role to fulfil in providing investment for new economic development. Scottish Homes is important in directing money towards housing associations and now area partnerships which deliver affordable housing. The spending commitments of the three water authorities are crucial if new infrastructure is to be provided where development is planned. There are two main problems at present. Firstly there is no requirement for these agencies to co-ordinate or comply with the development plan and secondly, development plans preparation timescales are too long for these organisations. In the field of transport there is no effective forward planning and co-ordination on a strategic level.

Historically, Regional Councils prepared both Structure Plans and Transport Policies and Programmes (TPPs), enabling the possibility of co-ordinating regional approaches to transport and land use planning. With the ending of TPPs and the transfer of responsibilities to smaller local authorities there is no longer a process for forward planning which enables regional priorities to be made. Authorities now have to bid for 'Challenge Funding' to the Scottish Office and compete against other authorities for transport investment.

With the exception of the Glasgow and Clyde Valley Structure Plan area, which lies within the boundaries of the Strathclyde Passenger Transport Authority, delivery of new public transport infrastructure is difficult. The strategic focus on transport investment has either moved to the private sector, in the case of the bus and rail companies, or to the Scottish Office in relation to roads investment.

Several solutions have been proposed to address the issue of delivering effective strategic planning: regional chambers which would integrate planning, transport and economic development and have involvement from MSPs, Councillors, Passenger Transport Executives and the LECs[11]; regional planning boards which would be professionally oriented with strategic planning powers[12]; and passenger transport authorities which would deal with the delivery of public transport. David Begg pursues these issues in Chapter Six. An alternative model would be the Government Offices approach currently operating in England where the Department of the Environment, Transport and the Regions devolves decisions on areas such as transport, planning and training to Integrated Regional Offices (IROs). This arrangement is likely to change as English Regional Assemblies are established.

These different models would require a level of decision making between the Scottish Parliament and the unitary authorities providing direct local authority influence, but avoiding parochial views dominating the process. Another solution would be to abandon strategic planning by local authorities altogether and to move to a system of national planning with regionally specific guidance, complemented by unitary plans prepared by each local authority.[13]

It is vital that the correct policy framework is set at national level with the institutional capacity to implement it. The establishment of the Scottish Parliament will enable national spatial planning, and should be backed up by regional planning strategies initiated by central government and involving local authorities and key stakeholders. This would focus attention on areas which are critical to Scotland's long term development (such as the central belt) which are currently dealt with by several structure plans.

The *process* of strategic planning also requires attention. Transparent, accountable and long term decisions about infrastructure, housing, economic development and transportation are likely to become progressively more difficult to achieve as the long term impact of local government reorganisation works its way through the system.

Changing Policy Context

Moving from *meeting demand* to *managing demand* has become a key objective in relation to transport and housing.[14] This is an important policy shift brought about by concerns about the cumulative impact of growth: 'The national forecasts of traffic are not targets that the Government is seeking to meet or to accommodate'[15]. There is now a growing consensus that the long term impact of development must be considered in order to avoid unsustainable outcomes which set a land use pattern for generations to come. Both these policy shifts were presided over by Conservative governments. In NPPG 3 Land for Housing (revised in 1996) the notion of local authorities being able to set the level of growth for their area was suggested, moderating the move to flexibility and consumer choice contained in the previous version of NPPG3 which had been published in 1993:

> The main change in this NPPG is to give greater weight to the importance of local considerations, balanced with national policy and other matters, when local authorities decide how and where to provide land for housing. [16]

The guidance also stresses the greater weight given to environmental considerations in comparison with previous guidance on housing. The Draft NPPG on Transport and Planning issued for consultation in 1996 set out the need for a more balanced approach to transport policy:

> Greater mobility has increased the number of journeys and their average length. The economic benefits offered by transport have to be weighed against the impact of transport infrastructure and traffic on land use, on air quality, on climate change and on quality of life in towns and cities. Land use policy represents one set of measures contributing to a balanced approach to transport policy.[17]

Retail policy is another example of a shift in government locational policy as a result of concerns about the growth of the number and length of trips in our everyday lives. There is now a sequential test applied to new retail developments:

> First preference should be for town centre sites, where suitable sites or buildings for conversion are available, followed by edge-of-centre sites, and only then by out-of-centre sites in locations that are or can be made accessible by a choice of off-centre sites in locations that are or can be made accessible by a choice of means of transport.[18]

Other issues such as the continuing expansion in leisure developments await attention. With new issues however, it could be argued that the planning system is less capable of anticipating change than responding after the event. The planning implications of technological change are particularly difficult to predict. Problems can arise at the local level in advance of a policy framework from either local or central government. The burgeoning telecommunications industry is a good example of the difficulties in providing a responsive policy framework.

New Issues Which the Scottish Parliament Could Address

These policy debates must be addressed regardless of the framework of government. However, the establishment of the Parliament offers the opportunity to look at some issues afresh. A review of the speeches, articles and pamphlets that have been written reveals a series of themes which ought to set the devolved planning agenda. Some of these focus on substantive policy ideas, others emphasise the means of implementation.

Sustainable development stands out as the main priority for the Parliament to grapple with. This will be a daunting challenge as more knowledge will be required in order to test the costs and benefits of different proposals. It will also involve conflict resolution, no simple task given the forces ranged for and against different types of development. There are methodologies such as Strategic Environmental Assessment which enable sustainable development issues to be considered as part of the forward planning process, in order to assess whether new plans are likely to lead to sustainable outcomes. However, the time constraints under which local authorities are operating means that new methods are difficult to incorporate into the process when the pressure is for faster and less complicated plans. In addition, these methods call for more rigour and better information for monitoring and measuring possible impact.

There is an important role for the Scottish Office to play in setting an agenda and promoting best practice in evaluating the impact and trade-offs that different decisions will have. The John Wheatley Centre Report on the environment (1997) recommended that all government departments should be required to audit their policy and financial decisions against sustainable development criteria. [19] State of the Environment Indicators would provide a focus at the national level on sustainable development issues which could inform overarching planning guidance and provide a resource base from which local indicators could be developed. A key conclusion from the research carried out by Friends of the Earth Scotland (1996) for their discussion paper *Towards a Sustainable Scotland* was that although obtaining up-to-date information is difficult, it is vital if sustainable development strategies and targets are to be established. [20]

There is an important gap to be filled around the need to develop national planning responses based on the fine grain of planning issues in Scotland. There has been a tendency (because Scotland has been considered as a region) to think about Scotland as a single planning entity when in reality there are significant differences in planning needs and demands in different parts of the country. During research carried out to deliver teaching and in discussion with fellow planners, the extent to which they regularly comment about the need for more advice specific to their area is striking. Rural Scotland has been increasingly covered in NPPGs that have been issued, for example in relation to skiing developments and coastal planning issues, yet planners will complain that rural transport issues are not effectively addressed in national policy. While there may be common agendas throughout rural Scotland there are also significant differences. Remote rural areas are seeking to tackle depopulation, whereas rural areas within travel to work areas for cities and major towns are experiencing immense pressure for development.

In urban Scotland the lack of specific advice and guidance on regeneration and particularly the need to set out a clear role for the planning system in delivering urban regeneration should be remedied. These issues are addressed in different NPPGs and PANs in passing. What is not provided is the specific regional context for policy development. A national planning strategy for Scotland would require these issues to be put into a national context and priorities and policies developed thereafter.

The need for a national spatial planning strategy is increasing advocated in the context of legislative devolution by commentators such as Hayton and Lyddon. This is an issue which has been resisted historically, possibly because of eastern European connotations. Lyddon suggests that the establishment of the Parliament should be used as an opportunity to reconsider the issue. He argues that Scotland could learn from the Dutch approach which enables a national vision to be set out and reflected in spatial priorities, but adopts the principle of subsidiarity and leaves local planning authorities to implement these priorities.[21]

More effective links between planning and urban regeneration should be forged. In the central belt of Scotland, how far sustainable development is approached will be strongly influenced by the extent to which the social, economic and environmental base of the cities is addressed. Focusing here will require change in the way the system operates. These have always been the planning system's objectives, but sustainable development requires a more informed and rigorous appraisal of developments.

West Lothian and Lanarkshire are two areas worth considering in this context. One of the challenges opened up by the recasting of strategic planning is that authorities which were formally part of one structure plan area are exploring different spatial perspectives. For example West Lothian District Council looked eastwards in the Lothian Region's Structure Plan. Now that there is no longer a Regional Council, West Lothian is in a position to capitalise on its position in the central belt on the M8 motorway. As well as looking to the east, there may be scope to improve links to the west by reconnecting the Bathgate rail link with Glasgow. Lanarkshire is strongly associated with the west central belt and the Glasgow conurbation. However, its focus could be widened by reconnecting with the east coast through its rail infrastructure.

In Glasgow and Edinburgh the extent to which the green belt is defensible in the long term is contested. Discussions focus on whether it would be sensible to leapfrog the green belt or to carry out strategic reviews leading to the release of 'less defensible' areas when appropriate development opportunities arise, as with the South East Wedge development around

Edinburgh. Expansion has tended to focus on adjacent areas and a gradual spreading of the cities has taken place as brownfield sites are developed. One drawback is that this tends to lead to dispersed residential development which is difficult to service with public transport resulting in people using their cars more frequently for commuting and leisure trips. One issue worth considering in the longer term is the possibility of linear expansion which can be serviced by heavy rail, providing new housing in less environmentally sensitive areas, with lower land costs. The new settlement trend has not been treated as a realistic option in central Scotland, but a form of efficient settlement expansion might be one means of absorbing residential demand that cannot be accommodated within existing city limits. This was considered in the English national planning advice document PPG 13 which focused on the need to reduce travel and dependence on the car. According to one recent analysis:

> Considerable uncertainties about the links between urban form and energy use in urban travel contrast strongly with the evidence of strong and straightforward effects from variations in fuel prices, which would be a much more reliable basis for an energy- and emissions-savings strategy. Higher prices would also tend to generate more compact settlement patterns.[22]

The use of economic instruments to complement the planning system has become an issue with the election of a different government. Road pricing mechanisms would make out-of-town locations which are not serviced by public transport less attractive to private car users, ensuring that where such activity takes place it at least raises revenue from users or businesses which can then be allocated to improving the public transport network. Specific examples of this type of mechanism would be charges/taxes for private non-residential parking spaces and charges for out-of-town supermarket/retail parking or for development of green belt sites for housing. While these are clearly options that could be pursued, there has been little evaluation of the potential impact of new charges and how they would interact with the system as a whole. There has certainly been no Scottish impact assessment to date on the housing markets or town and city centres. Adoption of road pricing would require a planning regime that did not create incentives for development to relocate to the area outwith the tolling area and lead to uncontrolled decentralisation. Levies on greenfield development would help to redress the balance towards urban regeneration, but again a clear planning framework would be required to identify preferred areas of development, with financial support for brownfield development, particularly where affordable housing is a priority or treatment is required before building starts.

Participation and Community Involvement

The John Wheatley Centre Paper on sustainability (1997) sees opportunities:

> ...for democratising the decision making process that currently underpins decisions about changes to our built and natural environment. The planning system holds out the hope that each individual has the right to be consulted about their future and the future of their community. In practice, this is a right many citizens know little, if anything about, and do not have the opportunity to exercise. It should be part of the mission of the Parliament to bring that vision about.[23]

There are a range of ways in which a more participative approach could be developed. National guidance could promote integrated decision making across housing, transport and economic development issues, It could also give a lead to local authorities, developers and community groups so that the criteria on which local decisions are based could become more transparent. All guidance could be subject to formal consultation to enable the genuine participation of stakeholders in advance of publication.

Keith Hayton suggests that this national focus on planning could take the form of a National Planning Forum which meets to discuss national planning issues and is mirrored by local fora.[24] The alternative approach would be parliamentary debate among MSPs and opinions invited from interest groups and communities in response to draft policy documents with an open public consultation process. The whole process should be transparent with all opinions and comments submitted subject to public scrutiny.

Information systems could also be developed to support the work of the Parliament. Ideas generated by the John Wheatley Centre on Telematics for the Scottish Parliament suggest a radically different way of operating[25] which could make the planning system more transparent and accessible. For example, as part of a national information resource, the national strategy, NPPGs and the digest of national planning information could be posted on the Scottish Office Website for planning authorities, businesses and individuals to access.. These documents could also be complemented by the availability of local planning authority documents. Some authorities are already exploring these possibilities and with wider access through schools and libraries these could increase knowledge of the system. Paul Filipek warns that 'at present there does not appear to be the degree of co-ordination and collaboration that is required if Scotland is to ensure that it keeps up with others in the emerging 'information society'. He suggests that:

> The effective employment of IT is becoming increasingly important in combating peripherality, achieving sustainable development and extending participatory democracy. The new Parliament must be ready to hit the ground running on this issue.[26]

Delivering ideas such as a national planning strategy and more geographically sensitive planning policies will require greater engagement from the Scottish Office in overseeing and shepherding the planning system. Derek Lyddon, former Chief Planner for Scotland, makes a plea for 'intelligent evolution rather than revolution' for the changes that will be made to the system.

Promotion of community involvement could take several forms and result in changes to planning law and practice:

- Community Councils and organisations could be given more influence in the planning process and equipped to play a more meaningful role. Local Agenda 21 networks could be important in thinking through new ways of implementing sustainable development.
- A nationally resourced Planning Aid Service, transforming the current limited service which exists, could ensure fairness in the decision making process by providing independent advice about the planning system, regardless of where people live. It would operate independently from both the Parliament and Local Authorities, but be funded by their contributions.
- Consideration could be given to promoting best practice techniques which would encourage more effective participation in development plan consultation so that ordinary people are able to be involved in shaping their future rather than merely reacting to it. The development of conflict resolution processes in planning inquiries offers one example where there are complex factors and a range of views which need to be taken into account.
- Third party right of appeal on planning applications could be applied where there are strong community objections or where a planning authority has not followed the policy framework set out in the development plan could be introduced.

If the planning system is to play its full part in promoting a more sustainable Scotland, innovative approaches could have an important role in problem-solving rather than encouraging stand-offs between opposing interests. Another important implication of moving towards sustainability will be a more rigorous consideration of the likely implications of new developments. Environmental impact assessment and strategic environmental assessment methodologies have been developed. The challenge is to see them adopted as a matter of course, not just after proposals have been drawn up but during the drafting of strategies and development plans as well.

Challenges Created by Devolution

There will also be challenges arising from the nature of the Parliament itself. The system of proportional representation will have a substantial impact on the balance of political representation - no one party is likely to exercise power on its own. The traditional electoral system of Westminster and local government ('winner takes all' under first-past-the-post) generally means that the ruling party is under no obligation to negotiate with the policy proposals of other parties. This may lead to discussions between parties in order to agree on specific policy areas where there might be consensus. There has generally been an absence of party political discussions about the planning system and its operation, although there may be different views on specific policy issues addressed by the parties. With more dispersed sources of political decision makers without previous experience of government, different approaches can be anticipated.

It will be interesting to see whether the MSPs are able to resist the temptation to centralise influence in the Parliament. The White Paper laid down a marker on this issue, stressing that:

> In establishing a Scottish Parliament to extend democratic accountability, the Government do not expect the Parliament and its Executive to accumulate a range of new functions at the centre which would more appropriately and efficiently be delivered by other bodies within Scotland.[27]

A change in relationships between quangos and government could also alter the terrain in which the planning system operates. The John Wheatley Centre's Quango Commission suggested different futures for quangos which currently influence the planning system: the policy making responsibilities of Scottish Homes should be the work of the Parliament and its research capacity might be carried out either by the civil service or by another independent research institution; the LECs' powers should be transferred to local authorities; and SEPA and SNH should remain at arms length from government in order to 'ensure that there is an efficient independent regulator which will have the ability, power and freedom to scrutinise the activities of the Scottish Government as well as local government and the private sector'.[28]

The greater relative weight of Scottish rural representatives in Holyrood compared to Westminster may change the nature and tenor of debate with particular implications for rural policy development in relation to planning issues. The possibility of urban/rural conflict and different perspectives on priorities for the planning system is likely to develop. An obvious example is the issue of continuing pressures on rural areas within commutable distance from the cities for new residential development. The recent suggestions for a Green Belt designation around St Andrews demonstrates that communities are already beginning to think more proactively about how the planning system affects them and how they might influence it to protect their perceived interests. The challenge for the planning system will be to consider the viewpoints of both protester and developer alike fairly and openly, and to be seen to do so.

One of the major issues identified in the 1997 referendum campaign was the equal representation of women in the new Parliament. If brought about this could also change the terms of debate with new policy priorities being developed. The Women and Planning movement has long argued for a broader perspective on land use planning issues which takes into account social issues in relation to planning. A more integrated planning perspective which considers locating residential, business, education and leisure facilities in closer proximity to public transport, walking and cycling routes, with a greater emphasis on public safety and planning could provide an initial agenda for women. In the longer term the profile that the Scandinavian parliaments (with much higher proportions of women) have given to their national planning strategies and policies could be a model to learn from.

Two final issues are explored here. The first is the potential divergence of planning systems within Britain. Although planning has been the subject of administrative devolution and has developed its own practice, there have been few significant differences between the Scottish and English planning systems. Scottish legislation on planning has tended to follow on from English legislation, the plan-led system stressed in the 1991 Planning and Compensation Act being a good example. However, devolution implies possible divergence both in legislation and practice. The need to discuss best practice throughout the UK will remain but take on a different focus.

The second issue is the relationship between the Parliament and the European Union. With the emergence of Europe 2000 and the European Spatial Development Perspective (ESDP), the need to broaden horizons and engage more effectively with European planning will become more urgent. While it is not the intention of the ESDP to be prescriptive, it clearly points towards a more powerful role for the European Union in setting the future agenda for involvement in planning and could lead to the development of physical planning priorities. It is clearly in Scotland's long term interests to play an active part in shaping European policy and ensuring that its concerns and priorities are reflected in the big picture. As Cliff Hague puts it:

> Engagement with spatial planning would help British planners to share ideas about sustainable cities with their European counterparts. Above all, a country in which so many regions are so peripheral form the core of European activity should be embracing a policy which can be made to focus on spatial integration.[29]

Notes

1 Quoted in *Review of Development Planning in Scotland*, Hillier Parker et al (1997), The Scottish Office.
2 Mackenzie A. (1997), The Northern Lights of Best Practice, *Planning Magazine*, 25th July 1997, pp. 16-17.
3 PAN40, June 1993, Development Control, Scottish Office Environment Department, pp. 5 & 8.
4 The Scottish Office Environment Department, *Review of the Town and Country Planning System*, 1994.
5 Hillier Parker et al 1997, *Review of Development Planning in Scotland: Interim Report for Consultation*, Commissioned for the Scottish Office Development Department, pp 15-17.
6 Hillier Parker et al, 1998, *Review of Development Planning in Scotland*, Commissioned for the Scottish Office Development Department, p35.
7 Calum Macdonald MP, February 1998, 'Planning for the Millennium', Speech at Conference on Planning for the Millennium, Edinburgh. See also Modernising Planning, 1997, A Policy Statement by the Minister for the Regions, Regeneration and Planning, Department of the Environment, Transport and the Regions, January 1998.
8 *Scotland's Parliament*, July 1997, Scottish Office, Cm 3658.
9 Hague C., 1997, 'In Search of the Scotland 2020 Vision', *Town and Country Planning Journal*.
10 Hillier Parker et al, 1998, *Review of Development Planning in Scotland*, Commissioned for the Scottish Office Development Department.
11 Begg D., 1997, *Integrating Transport, Planning and Economic Development : A Decentralised Model for a Scottish Parliament*, Scottish Local Government Information Unit.
12 RTPI, 1994, Briefing Note on the Local Government Etc (Scotland) Bill for the House of Lords Committee, July 1994.
13 Hayton K., 1997, 'Opportunity Knocks in Scotland', *Planning Magazine*, 11th July, p9.
14 SODD, NPPG3, Revised 1996, Land for Housing, November 1996, Scottish Office.

15 Draft NPPG: Transport and Planning, Scottish Office Development Department, May 1996, p7.
16 NPPG3, Revised 1996, Land for Housing, SODD
17 NPPG: Transport and Planning, Scottish Office Development Department, May 1996, p7
18 NPPG8, April 1996, Retailing, SODD.
19 John Wheatley Centre, 1997, *Working for Sustainability: Environment Policies for a Scottish Parliament*, JWC, Edinburgh.
20 FOE Scotland, *Towards a Sustainable Scotland: A Discussion Paper*, March 1996.
21 Lyddon D., 1997, 'Scope, Selection and Staff', *Town and Country Planning*, November 1997, Vol. 66, No.11, pp307.
22 Gordon I., 1997, 'Densities, Urban Form and Travel Behaviour', *Town and Country Planning*, September 1997, Vol.66, No.9, pp241
23 John Wheatley Centre, *Working for Sustainability: Environment Policies for a Scottish Parliament*, JWC, Edinburgh.
24 Hayton K., 1997, *The W(h)ithering of Scottish Development Planning? The Impact of a Scottish Parliament upon Development Planning*, Strathclyde Papers in Planning, No.31, University of Strathclyde, Department of Environmental Planning, August 1997.
25 John Wheatley Centre, 1997, *A Parliament for the Millennium: A Report by the Advisory Committee on Telematics for the Scottish Parliament*, July 1997.
26 Filipek P., 1997,' Statutory Strategy', *Town and Country Planning*, November 1997, Vol 66, No.11, pp306.
27 *Scotland's Parliament*, July 1997, Scottish Office, Cm 3658, p19.
28 John Wheatley Centre, 1997, *Quangos: Policy proposals for a Scottish Parliament*, Draft report, JWC, Edinburgh.
29 Hague C., 1996, 'Spatial Planning in Britain: the issue for planning in Britain', *Town Planning Review*, Vol 67, Number 4, October 1996, ppvi.

8 Agriculture, Forestry and Rural Land Use

ANDREW RAVEN

Introduction

As much as 98 per cent of Scotland's territory can be classified as rural, home to one-third of Scots. Rural Scotland provides key assets for a sustainable future: economic opportunities, cultural traditions and a rich environment. Any consideration of prospects for sustainability requires an understanding of the different rural land use sectors, the role of public policy and options for change. It is as important to grasp the overall picture as dwell on any particular aspect in great depth. This chapter therefore presents an overview of land use in rural Scotland.

The Scottish countryside occupies a unique place in the heart of Scots and the advent of the Parliament will create a new focus for Scottish rural affairs. In addition, the electoral system should ensure effective representation from outwith the Central Belt. Within this context, prospects for sustainability in rural land use can be divided into an analysis of the current situation (Audit), what we might want (Vision) and how we might get there (Action). Excluded from this analysis are marine and fisheries issues, which demand more specialist coverage and, although a significant part of the wider rural picture, are outwith the scope of this chapter.

Audit

Much of Scotland's rural economy is shaped by state intervention. Land use examples include dependence of agriculture on the Common Agriculture Policy (CAP) and of forestry on woodland grants. Over 20 per cent of the land surface is covered by one or more of the main conservation designations, whether for landscape or wildlife. Similar interactions between government and rural society span most rural sectors, including housing, transport and the provision of all services. There are overwhelming economic, social and environmental reasons for such support to continue and indeed be enhanced. Government through its various agencies is thus the key player in rural areas, yet this is too seldom appreciated and far from completely understood. The starting point for any review of rural land use policies should therefore be an analysis of existing policies and their benefits, disadvantages and interactions. In order of economic importance, the main extensive land uses include agriculture, forestry, conservation and field sports.

Agriculture

Farming remains the dominant rural land use. With varying degrees of intensity, it occupies about three quarters of Scotland's land surface of which 86 per cent in the uplands and hills is classified as 'Less Favoured Areas' (being predominantly suited to extensive livestock production). The gross output of Scottish agriculture was £1.8 billion in 1995, although the profit of £450 million was only roughly equivalent to the total annual direct subsidy. Overall direct subsidy in 1997 rose to some £480 million and it is estimated that agriculture benefits from the same amount again in indirect subsidies. Three pounds in every five (61 per cent) spent on subsidies are for crops, one-third is for livestock and only 3 per cent is for environmental measures. Employment in agriculture totals some 68,000 people, less than 2 per cent of Scottish jobs, split between farmers and employees. However in remote areas with few alternatives, up to 10 per cent of the work force may be involved in agriculture. The consistent trend has been for farming to become increasingly capital intensive, while shedding labour.

The dominant force in Scotland's agricultural economy is the European Union's CAP, which provides three-quarters of agricultural subsidies. This encompasses price guarantees for agricultural commodities together with area and stock headage payments, supply side limitations (quotas), and agri-environmental incentives. Brussels sets prices and subsidies in European Currency Units, so support in member states varies with currency fluctuations. Britain's departure from the Exchange Rate Mechanism was referred to in farming circles as 'Golden Wednesday' but the subsequent strength of the pound has contributed to increasing agricultural pessimism.

Scottish farms are nearly ten times as large as the EU average. This contributes to 80 per cent of financial support to farming being paid to only 20 per cent of farmers. British policy has resisted any attempt to cap support by linking eligibility to farm size (so called 'modulation', see the following chapter in this volume by Hugh Raven). There is strong evidence that farm subsidies are rolled over into inflated prices for land. Farming is a way of life and, with money in their pockets, farmers tend to compete for extra acres.

The CAP takes up over half of the total EU budget. Reform is the subject of almost continuous debate, normally linked to the accession of additional countries, reform of structural funds and world trade negotiations. The complexities of agricultural policy result in increasing burdens of bureaucracy on individual farmers: every field now has a unique map code number, every cow its own passport. Much of the cost of CAP is met by consumers, who pay food prices inflated above world market levels. Consumer concerns are compounded by food safety scares. Recent headline problems include salmonella, BSE and E.Coli. Genetic modification of crops is now common place. Cloning of animals is technically feasible. The conflict of interest between supporting producers and safeguarding consumers within one government department has at last been recognised with the establishment of a Food Standards Agency.

Crofting

Crofting is a unique form of landholding in the Highlands and Islands, introduced in 1888 as a belated response to the infamous Highland Clearances. Crofts are smallholdings, normally with access to more extensive shared grazings, which benefit from intergenerational security of tenure at very low rents (based on the unimproved value of the ground). There are some 17,700 registered crofts covering about 20 per cent of the total land area of the Highlands and Islands. Crofting provides a model for integrating social and environmental values with agricultural activities. Traditionally crofting supplements part-time employment elsewhere. While the diversity of agricultural activity undertaken has sadly declined, crofting remains labour intensive (and capital extensive), thus maintaining a rich environment and substantial rural population in some of the least productive agricultural areas of Europe.

Woodlands

Trees cover some 15 per cent of Scotland's land surface, the majority being the result of upland afforestation earlier this century with imported conifer species. Only one tenth of Scotland's woodlands are semi-natural, our native forest being reduced to only 4 per cent of its original estimated extent. Native woodland tends to be the climax natural habitat, with far greater diversity of dependent species than in those land uses influenced by human development. As areas planted this century become productive, output from Scotland's forests has more than doubled in the last 20 years and is predicted to do the same again in the coming 20 years. This has stimulated the development of processing industries for timber and wood pulp.

Forestry directly provides 10,000 jobs but is becoming polarised between commercial and socio-environmental objectives. The forest industry is increasingly dominated by short rotations of exotic conifers, using specialist machinery and itinerant labour. Native woodlands grow more slowly but provide a wider range of public benefits. The woodland sector is also divided between public and private, with extensive state forests, coupled with regulation and incentives to control and encourage private woodlands. Any extension of native woodlands is likely to continue to require positive discrimination. There is increasing evidence of woodlands offering opportunities for community involvement in natural resource management.

Natural heritage

The term natural heritage includes nature conservation, natural beauty and amenity. Scottish Natural Heritage is charged with securing its conservation and enhancement, fostering understanding and facilitating enjoyment of it. An indication of the absence of political priority afforded these aims was a budgetary cut of 10 per cent in 1996/97 and standstill budgets since in cash terms. This is despite environmental charities having over half a million members in Scotland.

The Government was elected in 1997 with a manifesto commitment to strengthen wildlife legislation and a preference for the devolved Parliament to introduce National Parks in Scotland. These will draw on international best practice, encouraging integrated and sustainable management. The EU Habitats Directive is also being implemented, designed to ensure that the country's finest nature conservation sites are maintained in favourable conservation status.

Despite a complex system of designations, biological diversity ('biodiversity') has continued its remorseless decline. This is partly because conservation has depended on the 'voluntary principle', whereby land managers are offered (frequently inadequate) incentives for sustainable management but with no comprehensive underpinning regulatory framework. Tensions also continue over rights of public access, particularly to unenclosed land, reinforced by prejudices amongst both landowners and walkers and by inadequate public funding. Painstakingly established policy consensus over sustainable development has been significantly damaged by the proposal to build a funicular railway for skiers on Cairngorm.

Green tourism offers a real opportunity to derive economic benefit from natural heritage qualities. The economic impact of mountaineering in the Highlands and Islands has been estimated at over £160 million per annum, making it one of the biggest industries. This has developed without external stimulation yet is critically dependent on the quality of the natural resource on which it is based.

Other land uses

Field sports are the dominant land use over extensive areas, especially in the Highlands. A mistaken belief that high deer numbers will yield better sport, coupled with the extension of favourable woodland habitat, has led to a doubling in numbers of deer over 30 years to currently unsustainable levels. Recent legislation governing the Deer Commission for Scotland included measures to counter this trend but they remain unused. Grouse shooting can provide significant economic benefits in remote areas but is notoriously fickle, exclusive and accompanied by persecution of raptors.

Tourism is the largest source of employment in many rural areas, with annual visitor expenditure of about £1.5 billion. Some 20 per cent of the gross domestic product of the Highlands and Islands and about 13 per cent of employment comes from tourism (Scottish Office, 1997). Mineral extraction can also be a significant source of wealth and employment in rural areas. Mines and quarries are slowly being brought up to modern environmental standards. Development of areas of our finest scenery as superquarries to supply unsustainable building programmes elsewhere is increasingly being questioned. Mineral extraction can never be reversed, so is inherently unsustainable. Reduction in demand and increased recycling would help reduce the impact.

Vision

There is remarkable consensus on the ultimate aims for rural policy but sharp divisions over how they should be achieved. The then Conservative Government defined its policies for the rural communities of Scotland in a White Paper (Scottish Office, 1995) entitled *Rural Scotland: People, Prosperity and Partnership*:

- We will work in partnership to enable rural Scotland to be:
- Economically prosperous, with a range of job opportunities which will enable those who live in rural communities, native or newcomer, to enjoy worthwhile ways of life.

- Vigorous in its community life supported by good local infrastructure and quality services.
- Culturally confident, cherishing local traditions and distinctive ways of life, and able to adapt to and benefit from changing circumstances.
- Able to protect, conserve and enhance its outstanding natural environment.

The newly-elected Labour administration produced a Discussion Paper in October 1997 entitled *Towards a Development Strategy for Rural Scotland* (Scottish Office, 1997) which reaffirmed these goals:

> The overall aim of all our policies for rural Scotland is to foster and enable the sustainable development of rural communities. Sustainable development is the overarching theme at the heart of all our policies...Sustainable development depends on taking an integrated approach to each of the three main policy objectives: economic, social and environmental. These three facets of sustainable development are equally important...The Government believe that rural development should be driven by the priorities of local people to a much larger extent than in the past. Rural communities should be able to shape their own future and take part in the decisions that affect the economic, social, cultural and environmental well-being of their area.

Few would disagree with this in principle, but what practical implications have to be considered?

Action

Turning rhetoric into reality involves consideration of both themes and sectors. The main themes for rural land use policy are integration, community empowerment and sustainability. Others as diverse as education and land reform are either beyond the scope of this chapter or dealt with elsewhere (see Chapter 9 for example). Key land use sectors covered here include agriculture, woodlands and natural heritage.

Integration

Rural public policy is plagued by sectoralism. Any one acre may be subject to incentives to plough, graze, or not graze, to drain or not drain, to plant with trees or keep scrub at bay, or zoned for or against developments. Different public bodies have remits focused on agriculture, woodlands, deer, natural heritage, cultural heritage, economic development, housing and the provision of services. Subdivision of government into narrow sectors fosters expertise and encourages transparency in trade-offs between the sectors but has been hamstrung by inabilities to comprehend the wider picture. The advantages of specialisation could be balanced by the desirability of integration through the development of strong partnerships.

A Scottish Rural Partnership, including local, national and funding components, was introduced by government in 1996 to provide coordination and resources. Initially it did not attract sufficient political priority to cut across existing sectoral agendas. The incoming government in 1997 committed itself to extending the work of the National Rural Partnership 'in promoting a shared vision of the overall aims for rural development in Scotland, and a more integrated approach to rural policy'. It also established an inter-Departmental Committee for Rural Affairs to co-ordinate rural policy, and initiated research into 'the contribution which different sectoral policies make to the overall aims of rural policy with a view to improving integration and the effectiveness of policy delivery.' These steps towards integration, partnership working and a unified interagency approach need to be consolidated and given enhanced priority.

Community empowerment

The United Kingdom has long recognised overseas, and is gradually acknowledging at home, the key Rio Principles that:

> Human beings are at the centre of concerns for sustainability...local communities have a vital role in environmental management and development...States should recognise and duly support their identity, culture and interest and enable their effective participation in the achievement of sustainable development (UNCED, 1992).

There is increasing evidence of Scottish communities taking effective control of local resources, but this is often *despite* rather than because of the system they face. Many agencies have internal structures and procedures that impede rather than liberate local initiative.

There is no single right answer. Grassroots involvement in developing and implementing rural policy should involve huge diversity. Communities must be offered a menu of options from which to choose solutions that best suit their individual circumstances. This will require facilitation by those skilled in rural development. An exemplar could be the Action Network developed by Rural Forum Scotland, designed to support and connect individual rural development field workers across all sectors with the aim of stimulating more appropriate and effective development. Membership is open to all individuals who are actively involved in their rural community in whatever capacity.

Sustainability

Scottish Natural Heritage have suggested that the core principles of sustainable development are wise use, carrying capacity, environmental quality, the precautionary principle and shared benefits (SNH, 1993). Whatever definition is adopted, no consensus has yet developed over the objectives or indicators of progress in relation to rural land use. Different sectors will give different weightings to the economic, social and environmental components of sustainability.

Underlying the whole debate about public policy and rural land use are unanswered questions about where the greatest public benefits lie. Is the countryside best used for primary production, for the maintenance of biodiversity or for public recreation? What constitutes good land management? There are many detailed codes of good practice but no consensus on the overarching context. Unless and until we develop accepted standards of management, much debate will continue to generate more heat than light. To be sustainable, ultimately the rural economy will need to internalise all those external costs currently borne by wider society. Rural land use offers key opportunities, for example in relation to conserving and enhancing biodiversity, or renewable energy generation, but also suffers from key threats, particularly from heavy reliance on fossil fuels.

Agriculture

Current agricultural policy is unsustainable politically, economically and environmentally. The policy was initially developed at a time of food shortages, driven by the concern to increase production and ensure stability of food supply in the public interest. Technological and managerial improvements by the early 1990s had resulted in substantial surplus European production of many farm products. Policy development since has aimed to curb costs and production but has yet to articulate an alternative public benefit as the core rationale for continuing expensive support.

Britain has and must continue to argue within the EU for social and environmental objectives replacing production as the basis for support. Reform of the CAP thus needs to move towards income support for public non-market benefits delivered. Environmental considerations should be a core objective of mainstream support, unlike at present where agri-environmental incentives are an under-funded afterthought. This should be accompanied by the principle of cross-compliance, ensuring that receipt of public subsidy is dependent upon maintaining public goods, thus encouraging positive environmental management. Reform should be designed to encourage rural employment, but change will also heighten the need for land managers to diversify away from over-reliance on agriculture.

The EU flirted with such reform in the so-called Cork Declaration of 1996 which suggested that sustainable rural development should be given priority and underpin all rural policies. Policies should be multidisciplinary and multi-sectoral, applying to all rural areas but targeted on those of greatest need. Policies should protect and sustain the quality and diversity of rural landscapes, being decentralised with a single programme for each region. However this progressive view failed to survive the political pressures within the EU and the agricultural lobby.

Woodlands

Britain relies hugely on imports for timber and forest products, while overproduction of agricultural commodities offers further opportunities for afforestation on agricultural ground. There is scope to extend woodland of whatever type, but this should increasingly come 'down the hill' from the uplands where artificial incentives have largely forced it to date.

Investment in woodlands is hampered by the very long wait for returns, when macro-economic trends place more emphasis on short payback periods. This gives a key role to government, who must establish and abide by long-term targets, supported by a suite of regulations and incentives. This should be accompanied locally by the continued development of integrated Indicative Forestry Strategies. Post-war forest policy was initially designed to provide jobs and homes in rural areas. These social benefits have been increasingly eroded and require reinforcement. Forestry has scope to be a key rural development activity, with increasing evidence of successful community involvement. Enhancing the environmental benefits of woodlands will also require positive discrimination, including targeting research on native woodlands and markets for native timber.

An example of the potential for integrating social and environmental development into woodland management is provided by the Millennium Forest for Scotland initiative. This is helping Scottish communities celebrate the Millennium through restoring and extending native woodland. It uses new money from the National Lottery to deliver grassroots projects in a cost-effective manner.

Natural heritage

Society increasingly recognises that the conservation and enhancement of natural processes is as important a public benefit as primary productive land uses. As non-market benefits, natural systems require positive discrimination and increased institutional support. Sustaining and improving biodiversity and public enjoyment of the countryside requires an increased proportion of public spending on rural land use. It also requires cross-sectoral support, including balancing duties for sustainable development being introduced into the statutory requirements of all agencies.

Designation of appropriate areas is internationally recognised as a key mechanism for protecting natural heritage values. The UK designation system has tended to be complex yet to have large gaps, relying on inadequate regulation rather than incentives for sustainable management. The result has been a system that is misunderstood and unpopular with land managers and produces continued attrition of natural heritage values. Reform is required to both strengthen regulations and improve incentives for appropriate management, supported by an integrated suite of designations within a comprehensible hierarchy. Differing balances of priorities between environmental, social and economic issues should be provided by zoning.

Other land uses

The themes of integration, community empowerment, sustainability and environmental education apply across the whole spectrum of rural land uses. It is increasingly appreciated that the management of freshwater requires to be integrated within whole catchments. Coastal management is increasingly integrated within appropriate zones. Deer management needs to move away from the current unfortunate conjunction of excessive numbers with damage to other land use interests, towards a new managing culture optimising economic, social and environmental benefits through holistic management. Even the proponents of mineral extraction would recognise the value of strategic planning of supply and demand, potentially avoiding divisive and adversarial Public Inquiries into applications for planning permission. An increasing emphasis on reuse and recycling could minimise damaging impacts.

Conclusion

Scotland lies on the north west fringe of Europe, distant from many centres of political power. Its landscape and climate contribute to some of the lowest population densities in the EU. These and many other factors result in longstanding concerns about rural decline, but political and environmental developments are increasingly resulting in such pessimism being replaced by an appreciation of the real opportunities available to rural Scotland.

Public policy is the key influence on rural land use. There is increasing consensus on the objectives of rural policy, but this has yet to be matched by consistent action. A brief but comprehensive code of good land use practice is essential. Without it, there is no basis on which to assess quality of management and the balance achieved between public and private interests. This should be accompanied by a clear trend towards the 'polluter pays' principle in rural resource management. Agricultural policy needs environmental and social benefits as the core purpose of ongoing support, accompanied by incentives to diversify. Woodland policy also needs a stable framework to encourage the integration of economic, social and environmental benefits. Maintenance and improvement of the natural heritage needs to be given at least equal status with primary production, accompanied by strengthened regulations and improved incentives for sustainable management practices. A devolved Scotland provides a unique chance to move towards more integrated and sustainable land management systems, aided by involved local communities.

References

Birdlife International (1997) *An agenda for action: reform of the CAP*.

Crofters Commission (1992) *Crofters & Crofting*, HMSO, Edinburgh.

FAPIRA (1995) *Forests and People in Rural Scotland: A Discussion Paper*.

Highlands and Islands Enterprise (1996) *The economic impacts of hillwalking, mountaineering and associated activities in the Highlands and Islands of Scotland*.

Hunter, J. (1991) *The Claim of Crofting: The Scottish Highlands & Islands 1930-1990*, Mainstream.

National Farmers Union of Scotland (1997) *Policy options for Scottish Agriculture: a Discussion Document*.

Native Woodlands Policy Forum Scotland (1996) *Native Woodlands and Forestry Policy in Scotland: A Discussion Paper*, WWF Scotland, Aberfeldy.

Scottish Natural Heritage (1993) *Sustainable Development and the Natural Heritage: the SNH approach*, Perth.

Scottish Natural Heritage (1994) *Red Deer and the natural heritage*: SNH Policy Paper, Perth.

Scottish Natural Heritage (1997) *Annual Report 1996/97*.

Scottish Office (1995) *Rural Scotland: People, Prosperity and Partnership: The Government's Policies for the Rural Communities of Scotland*, HMSO, Edinburgh.

Scottish Office (1997) *Towards a Development Strategy for Rural Scotland: A Discussion Paper*, The Stationery Office, Edinburgh.

Scottish Office (1998) *Scottish Agricultural Income Estimates 1997*, News Release 0159/98.

The United Nations Conference on Environment and Development (1992) *Earth Summit,* The Regency Press Corporation, Rio de Janeiro.

John Wheatley Centre (1997) *Working for Sustainability: An Environmental Agenda for a Scottish Parliament*, John Wheatley Centre, Edinburgh.

9 Land Reform

HUGH RAVEN

Introduction

This chapter presents a brief review of the debate on land reform in Scotland. It does not contain new material but instead draws upon the main recent publications on the subject, which deserve credit for helping to sketch out a programme of land reform for the Scottish Parliament. That process took a major step forward with the publication in February 1998 of the first report of the Scottish Office's Land Reform Policy Group (henceforth referred to as *Identifying the Problems).*[1] The Policy Group's terms of reference are:

> To identify and assess proposals for land reform in rural Scotland, taking account of their cost, legislative and administrative implications and their likely impact on the social and economic development of rural communities and on the natural heritage.

The report is a clear and concise digest of the most pressing problems and anomalies in Scottish rural land law, and brings a very welcome spirit of consultation and openness.

For historical and economic reasons, which together mean Scottish land ownership remains highly concentrated, land reform is a more significant issue in Scotland than in other parts of the UK, and probably any other European country. But the issue is also significant because land-based employment in Scotland is more important as a proportion of total employment than the rest of the UK, and because it is governed by a body of law at once the least changed since it originated almost a millennium ago and the most distinct from that applying in the rest of the UK.[2] Wightman (1996) has defined land reform in a Scottish context as:

> A comprehensive package of legal, administrative and fiscal measures designed to redefine and redistribute property rights. It contains many potential elements which in the Scottish situation could range from the reform or abolition of feudal tenure to the reform of the Common Agricultural Policy.

That is a long agenda, and much of it will be outside the remit of a Scottish Parliament.

Incentives for good management

One of the most important environmental aspects is the availability of financial incentives for good land management. In the short term the bulk of finance available, in the form of agricultural subsidies, will be least under Scottish control.[3] That may partially change, in that official proposals for reforming the Common Agricultural Policy may give more discretion to EU member states though it is too early to predict the outcome of the CAP reform process.[4] In the meantime the remaining elements of the land management budget (those agricultural payments under national control, forestry grants, and incentives for amenity or conservation management) will be controlled within Scotland and will give national government significant influence over land use.

Tenure

Arguably the most important influence over land use which will be almost wholly within the jurisdiction of the Scottish Parliament is land tenure. Tenure is a critical (though largely unquantifiable) determinant of management and, in turn, of environmental impact. This chapter therefore concentrates on some of the proposals for changes in tenurial arrangements affecting rural land.

Reforms: Proposed and Underway

A national land register

Despite the level of public interest in land ownership and use, there has been no comprehensive and readily comprehensible official source of information on who owns Scotland.[5] The gap has been largely filled by the publication of Andy Wightman's (1996) *Who owns Scotland?* (rumoured to have become required reading among the land-owning classes). But for a more detailed picture, the availability of information depends on a combination of historical circumstances - meaning that in some localities the land ownership pattern is easy to discern, while in others it is virtually impossible.

Arrangements are in hand to transfer much of the publicly-available information from the Register of Sasines (the national record of title deeds to land which has changed hands) to a computerised system. In a separate pilot scheme covering certain areas, these records and related details (such as topographical and land use information) are being transferred to a computer-based and accessible public register. This should cover the whole of Scotland by 2003, but will have considerable gaps where ownership has not changed for many years. These could only be fully addressed at very large cost, by requiring the compulsory registration of all titles and providing the necessary administrative infrastructure. Registration of all land holdings above a certain size could be much cheaper, the ultimate cost depending on the threshold chosen.[6]

Subsidies for land management

While ownership is usually a central determinant of land uses, so too is financial incentive. Previous British governments' obsession with secrecy has meant that although we know that four UK land managers have received more than half a million pounds in annual agricultural subsidy (and one over £1 million), the public is not allowed to know who they are on grounds of 'commercial confidentiality'. There is a different attitude in France, where lists of such payments are pinned to parish noticeboards. *Identifying the Problems* asks whether the same should apply in Scotland.

Feudalism

It is important to distinguish clearly between the meaning of the term 'feudal' in the narrow legal sense, and its wider connotation of an archaic social hierarchy (which may or may not accurately describe social relations in much of rural Scotland). Feudalism as it survives in the legal sense means that when land is sold, the vendor can impose conditions on its future use. Much Scottish land has feudal conditions attached, allowing previous owners to demand payment if the current owner wants the conditions to be relaxed, a system described as 'lucrative legalised blackmail'. *Identifying the Problems* makes clear that feudalism will be abolished as the minimum requirement of land reform, in line with a commitment from each of Scotland's main political parties. While this would enhance the legal rights of land ownership, contrary to widespread public misunderstanding it would not affect the distribution of land holdings.[7]

Rates

Agricultural land has always been exempt from non-domestic rates. The exemption was extended in the early 1990s to sporting rights, when landowners hoodwinked government into believing that the money would be better spent locally on conservation. In some areas there have been valuable initiatives as a result but in others there is no sign of any conservation improvements, and land degradation by excessive deer populations continues. Public pressure to re-impose rates is growing, partly as local authorities could ensure that the money so raised was indeed spent on conservation.

Land Transfers

Market regulation

Most land reform proposals include a mechanism for intervening in the market in the public interest. Such a power to cover all land sales would require compulsory notification of sales to a public body, as has been specifically proposed by Wightman.[8] Government has the power to regulate sales of private companies to prevent monopolies. Powers to intervene in the land market would follow the same principle. A land commission could be established with powers analogous to those of the Monopolies and Mergers Commission, substituting social and environmental criteria for the MMC's considerations of competition.[9] There are numerous suggestions as to what those criteria should be, considerations which would effectively be conditions of eligibility to purchase land.[10]

Conditions of eligibility

Purchase of land could be made conditional on the preparation of a satisfactory management plan.[11] A residency condition could be applied,[12] such that owners had to live on or within a fixed distance of their property. Conceptually similar is the proposal that absentee landowners be obliged to nominate a legal proxy or representative, resident on or within reasonable distance of the property.[13] There are also suggestions that certain limited categories of potential purchaser be prohibited from owning land,[14] or that a legal and financial link be required of landowners with the public authorities in Scotland,[15] to allow the application of sanctions for anti-social land management. To encourage wider land ownership, size limits could be imposed on the land holdings of any individual.[16]

Expanding Social Ownership

Community ownership and new settlements

Interest in expanding existing types of community ownership and investigating new models has increased enormously in recent years, fuelled by high profile community buy-outs (including the Assynt crofters and islanders of Eigg); James Hunter's work, including *inter alia* his paper commissioned by the then Scottish Secretary Michael Forsyth MP[17]; and the creation of the Community Land Unit in Highland and Islands Enterprise at the instigation of Brian Wilson MP. *Identifying the Problems* specifically covers the possibilities of new crofting communities and compulsory purchase of badly-managed land, with local community involvement.

For crofting communities the opportunities for community ownership were significantly increased by the Transfer of Crofting Act 1997, allowing tenants of the Secretary of State's 53 crofting estates to take their land into community ownership at nominal or zero cost.[18] The much larger number of crofters on private land are able to buy their crofts individually[19] and could theoretically adapt the procedure to allow community buy-outs of privately-owned croft land. New legislation to simplify this procedure is mooted in *Identifying the Problems*.

Many crofts were created in the early decades of the 20th century on land bought from private owners. Interest in the creation of new holdings and new communities (which would not necessarily include crofts) has reached a level not seen since before the last war, on the back of rising demand for vacant crofts and population growth in the Highlands and Islands.

New holdings could be created on land sold voluntarily, or by use of the extant legislation enabling land settlement, although the process would be complex and the legislation has not been used since the 1950s nor used extensively for 30 years before that. Any new settlement, voluntarily or compulsorily acquired, would involve substantial cost including the capital cost of acquiring the land and the continuing support for crofters or smallholders in the form of crofting grants and livestock subsidies.

Consistent with the principle of transferring state-owned assets to community control, as in the 1997 Act on crofting transfers, is local community control over Forestry Commission land. Several such proposals are under discussion, involving the transfer of either ownership or of management responsibility to the community. The best-known example is at Laggan in Inverness-shire where a community trust with a trading company subsidiary manages the land.[20] The same model is being explored elsewhere and there is a suggestion that in addition to woodlands it could apply to national nature reserves, or at least those areas on the periphery of reserves which are themselves of lesser conservation significance.[21]

Voluntary sector ownership

An important and high-profile change in Scottish landownership is the increasing significance of not-for-profit organisations. These now own 2.5 per cent of Scotland, a proportion which is increasing fast.[22] Many non-governmental organisations are bound by charitable objectives. While this may reassure the public of these organisations' *bona fides*, it can also lead to controversial management objectives as, for example, where they have a charitable purpose which limits their interest to species other than humans.[23] Given the accountability of membership NGOs and their public prominence, existing NGO landowners are at pains to adopt best practice and some, such as the John Muir Trust, involve their crofting tenants in management decisions which affect their interests. To promote this stakeholder model, there have been suggestions that NGO access to public funds for land purchase might be conditional on partnerships with local community groups, to ensure that social objectives are not overlooked.[24]

Tenanted land

The large-scale transfer of formerly public assets to private owners (some under voluntary arrangements including sales of Forestry Commission land and some under right-to-buy legislation for crofters, as above, and for local authority tenants) has left tenants of privately-owned land at an apparent disadvantage. There have been several proposals fort right-to-buy opportunities to be extended to agricultural tenants of private land.[25] Given the shortage of land for housing and commercial development in some parts of Scotland, and the reluctance of local authorities to exercise their compulsory purchase powers, there have also been suggestions that land be compulsorily leased on a long-term basis with rents related to its undeveloped value.[26]

Suggestions for the way Ahead

Information

The detailed debate about land reform has some characteristics in common with the Free Church: a small group of adherents beset with theological disputes. One condition for progress on land reform is greater public involvement - a process which has recently taken a huge step forward with the publication of *Identifying the Problems*. Another is the availability of more and better information about existing patterns of land use and tenure. Professor Bryan MacGregor argued in the first John McEwen Memorial Lecture in 1993:

> Land tenure should be a legitimate and important area of study for those with an interest in the outcomes of land policy. My contention is that land tenure can be studied rationally and that such study is central to an understanding of the current and future effectiveness of land policy. Given the vast public investment in rural land use, it remains an under-researched topic.[27]

Without further information, and without a systematic means of gauging the public's priorities, any list of suggested reforms is bound to be subjective and speculative. More information would help resolve a further distortion, namely that the debate has tended to concentrate on land which is being sold, for the obvious reason that an acute threat to a community is newsworthy in a way that chronic bad management is not. This has been at the expense of thinking about the majority of land which is *not* changing hands. *Identifying the Problems* helpfully covers issues affecting all rural land since all land is relevant.

Land ownership is a subject of considerable public interest and lavish public expenditure. There is an overwhelming case for details of ownership to be readily available. A national register, readily accessible to the public, is therefore a minimum requirement. The speed with which this happens, and the degree of detail covered, is a decision for the Scottish Office, in balancing the use of funds between providing public information and direct encouragement of beneficial land management. Certainly if the Scottish Office estimate is accurate, the cost of at least £300 million to provide a comprehensive public register, detailing every title in Scotland, cannot be a high priority. But information on all holdings of above 1000 acres, as the threshold for compulsory registration mooted in *Identifying the Problems*, should satisfy most enquirers and should be a high priority. The same openness should also apply to public subsidies. There is no justification for the identity of recipients of public funds to remain secret. Those who benefit from Local Enterprise Company and forestry grants are already identified. So too should be those who receive agricultural subsidies.

Tenure

Abolition of the feudal system is now not in doubt, though what will replace it is an open question (and one so technical and abstruse that it is beyond the scope of this chapter).

Market intervention

Proposals to allow public authorities to intervene in the sale of land need not introduce any new principles. Powers are already widely available to public authorities which compromise property rights in the public interest: controls under competition policy on company sales; the town and country planning system; and existing powers of compulsory purchase. These powers are not controversial. However, if they were to be extended to the type of cases that have recently received publicity, new criteria for intervention would be required.

Eligibility conditions

An obligation on a purchaser to produce a satisfactory management plan gives rise to questions about how such plans would be assessed, and, more controversially, how their provisions would be enforced. Enforcement powers are available to local authorities for breaches of planning restrictions, and it would be quite logical to extend them to cover land-uses which are currently exempt. But it is hard to see how they could cover social criteria, without a huge bureaucracy, complexity and a degree of subjectivity which would engender little public confidence. Similar powers do exist for the removal of agricultural subsidies in cases where conditions are flouted (such as avoiding over-grazing). But they are very seldom used because abuses are so difficult to prove. Social criteria would be even more so.

A residency condition for potential purchasers could present problems where there are planning restrictions or physical remoteness, and there may be difficulties in defining residency[28], although with flexible interpretation they should not be insuperable. Such a condition would however, imply a causal connection between absenteeism and bad management. While this is certainly the received wisdom, it would be rash to impose such a restriction without further research and evidence (and *Identifying the Problems* suggests that some absentee landowners have a better record than other resident ones). The same would apply to proscribing certain categories of buyer. Specific types of company are often cited as the worst category of absentee land-owner, but there is little objective evidence on which to base a firm judgement, and a ban on ownership would be administratively complex. If applied to existing land holdings, it could also be unlawful under the European Convention on Human Rights.[29]

Limits on land-holdings

An upper limit on the amount of land which can be owned by one person would be largely meaningless unless it also considered productive potential. Land is already classified by productive quality, but the categories are very broad and the process of allocation inevitably involves a degree of subjectivity. So the scope for disputes and the use of arbitration machinery could well become unmanageable. Such limits could almost certainly be circumvented by the use of nominees and subsidiary companies.

If size of land holdings is thought to be a major issue (and there is evidence in the agricultural sector that smaller units employ more labour per unit area), then a more productive method would be limiting the amount of public subsidy per business, or tapering it above a certain level (a process known as 'modulation'). Public expenditure is easier for public authorities to control than title, and the principle of modulating payments is already enshrined in some agricultural subsidies, such as beef headage and organic farming conversion payments.

Community Ownership

An expansion of community ownership will require compulsory purchase; significant new sources of money for voluntary acquisition; or transfers of publicly-owned land. All three are suggested as possible, at least by implication, in *Identifying the Problems*.

Compulsory purchase

The use of compulsory purchase is presented by the Scottish Office as a power of last resort: 'wherever possible other mechanisms are to be preferred.'[30] However, the threat implied by such powers could be used much more effectively to improve land management; and their carefully-targeted use could be an effective combination for increasing community control. For example, many land-holdings include strategically-important assets, such as points of access or productive resources. Compulsory purchase of these would devalue the remainder of the holding, bringing it within reach of a larger number of potential buyers.

The use of compulsory purchase is well-established, and consistent with widely-accepted legal and moral principles. And there is widespread dismay at the reluctance of public authorities to use it, even in particularly flagrant cases of mis-management or abuse of monopoly ownership. If the Scottish Parliament is to be serious about land management in the public interest, compulsory purchase powers must be a major component of land reform.

Funding

The Heritage Lottery Fund is able to support land acquisition, though controversially it has chosen not to in several cases in the Highlands (and its priorities are changing, with new legislation diluting its land purchase priorities). There has been an increase in public funding to assist community control with grants from the Millennium Commission, though there is a strong sense that their interventions have been non-strategic and *ad hoc,* and require greater consistency and transparency in the choice of beneficiary.[31]

The Community Land Unit in Highlands and Islands Enterprise (HIE) has a modest land budget, and has contributed 'the last brick in the wall' to funding packages. While there are no strict guidelines on this contribution, it has rarely exceeded 10 per cent of the purchase price. Most of this will come from other existing budget lines. The recent half-million pound acquisition of Orbost on Skye in partnership with Skye and Lochalsh Enterprise (the LEC) was as large a project as HIE is able to manage, and has exhausted its land purchase budget. The resettlement of Orbost is a unique pilot project in rural regeneration, and is unlikely to be repeated until its success has been evaluated. The Eigg Trust too contains a strong element of experimentation. If these cases are to be repeated elsewhere in Scotland, there will have to be clear criteria for assessing public demand for new settlements. In Eigg, such demand was clearly evident, as was the capacity of the local community to manage the land once acquired. In other areas it may not be - in which case an objective means of testing whether they exist is required.

Cross-compliance

Another mechanism for increasing opportunities for local communities is the idea of making grants to existing landowners conditional on their making land available (where there is proven demand) for new settlements, community facilities or commercial activities. This could serve the public interest well, particularly where land is under monopoly control. In the hypothetical example often cited of a landowner approaching a Local Enterprise Company for a business grant, such a scheme could have the double benefit of releasing land for more commercial activity and providing the landowner with capital from sale of the land in question in lieu of some of the public sector grant. It is a principle called conditionality or cross-compliance that is becoming well-established in agriculture.

Transfer of publicly-owned assets

The same applies to a transfer to community control of Government assets, be they Forestry Commission sites or nature reserves: community interest and capacity for management must first be established. The case for such transfers is strengthened by the Forestry Commission having been forced to abandon its traditional role of providing long-term local employment, in favour of using mainly non-local contract labour squads. The sale or transfer of Commission plantations would be consistent both with the Commission's own historical purpose and the more recent policy of disposing of plantations into the private sector. But existing community initiatives have fallen foul of the Treasury's insistence on full market value for the transfer of state assets, and the Commission's own targets - set by government - for receipts from disposal of property. If under the devolution legislation the Scottish Parliament is vested with powers over Forestry Enterprise assets, a series of transfers to community control must be a priority.[32]

The right to buy

Extending the right-to-buy to private tenants of agricultural land is sometimes advocated on the grounds that it will bring new entrants to farming. But its effect would probably be exactly the opposite, as landowners bring more land in hand and decide against creating new tenancies. Given the controversy over similar (though short-lived) proposals for housing from the Labour Party when in opposition, such a reform seems unlikely: it is one thing for the state to expropriate other public authorities of their assets (such as with council house sales); it is quite another retrospectively to change a contract, freely entered into, and allow expropriation of private property.

One additional point merits mention. Much of the land reform debate is suggestive of a holy grail of land ownership, an optimal model of tenure which should be replicated as widely as possible in the public interest. Such a model does not exist. A diversity of tenurial options is desirable, including community, voluntary, private and public. Only a pluralistic solution can engage the interest and benefit from the talents of the maximum number of people.

Conclusion: A Land Reform Agenda for the Scottish Parliament

Information

As already observed, the land reform debate in Scotland has been driven by a number of exceptional, and sometimes exceptionable circumstances. But there is a dearth of hard information. Immediate priorities for the Parliament should be:

i) the preparation of a national register of land holdings above a certain level;

ii) publication of the details of all public land use incentive payments, including agricultural subsidies; and

iii) establishment of a systematic and long-term research programme into land ownership and management.

Tenure

With better information about existing land holdings and uses, it will be easier to judge the suitability of some of the proposals to proscribe certain types of landowner, or impose conditions on new purchasers of land. More immediately, the Scottish Parliament should:

iv) abolish feudal tenure;

v) establish strategic criteria for the greater use of compulsory purchase, the use of other public funds (such as Heritage Lottery Funding) and the transfer of publicly-owned assets to communities. It will be important to establish means for assessing:

- the quality of land management, to identify chronic mis-management;
- public demand for new land for community development (for housing, crofting, community facilities or forestry for example); and
- the capacity of beneficiaries to manage such land once acquired.

Incentives

During its early years the Parliament is unlikely to have significant amounts of new money at its disposal to affect land management directly through land acquisition. But it will be able to influence management through:

vi) developing and applying a clear set of environmental conditions for the receipt of public funds under the Parliament's control. In the short term this will include forestry and amenity planting grants, and land management payments through Scottish Natural Heritage. In the longer term, moves to greater subsidiarity may extend this to include discretion over CAP agricultural subsidies;

vii) modulating payments by tapering them above a certain level, linked not to the size of the holding but the amount of employment; and

viii) re-imposing rates on sporting land, and ear-marking the revenue to encourage conservation through local partnerships (for example deer management groups or fisheries trusts).

Notes

1 The Scottish Office Land Reform Policy Group, *Identifying the Problems*, February 1998, HMSO, Edinburgh.

2 Callander, R.F., *The System of Land Tenure in Scotland* - A Review, 1997, WWF, Aberfeldy. Ironically, however, some aspects of Scots law are sometimes considered preferable to their English and Welsh equivalents - as with the method of bidding for property, currently being examined with a view to replicating it south of the border.

3 About three-quarters of agricultural subsidies - which account for the vast majority of public support to Scotland's land-based industries - are allocated centrally by the EU.
4 Commission of the European Communities, 'Agenda 2000', Brussels, 1997.
5 See Wightman, A *Who Owns Scotland?* Canongate, Edinburgh, 1996 - who devotes a chapter to outlining the difficulties of obtaining the information which his book so valuably provides.
6 A threshold of 1,000 acres has been suggested by the Scottish Office in *Identifying the Problems*, which would cost £5 to 15 million - which could fall on the exchequer or on the title-holders.
7 This confusion is surprisingly widespread - including, for example, in an editorial in *The Herald* on
land reform (26th April 1996).
8 Wightman, A, 1996 - p211
9 MacDonald, Calum, MP, in Hunter, James, 'Progressing Community Ownership and Promoting New Types of Rural Settlement', Convention of the Highlands and Islands, 1997.
10 Logic dictates that such conditions should also apply to existing landowners - though legislative changes are rarely retroactive.
11 See Hunter, J., 1997.
12 As proposed by Wightman.
13 Scottish Land Commission, 1996.
14 See Wightman, p208, who suggests certain types of UK trusts and companies and offshore and foreign companies be proscribed.
15 Scottish Land Commission, 1996.
16 As proposed by Wightman
17 Hunter, J. 1997.
18 Though this risks over-simplification, given current negotiations over mineral rights.
19 Under the right-to-buy provisions of the Crofting Reform (Scotland) Act 1976: see Hunter, James, 1997.
20 Bids for community *ownership* have been stalled by the Treasury's veto on state-owned assets being transferred at less than full market value.
21 See Hunter, J., 1997.
22 See the Governance of Scotland Project, 1997; and Wightman, A, 1996.
23 The Scottish Land Commission commented of the National Trust and the RSPB that 'the extent of land owned by these two bodies alone ... is beginning to reach a stage when the remedy is almost as bad as the disease.'
24 See Hunter, J, 1997.
25 Hunter, J, 1997, suggests this be applied to all crofters; Wightman, A, 1996, and Scottish Land
Commission, 1996, both suggest it be applied to all agricultural tenants.
26 Hunter, James, 1997.
27 MacGregor, Prof. Bryan, 1993.

28 Although a definition applies in the tax system.

29 See The Scottish Office Land Reform Policy Group, ibid, para. 3.3.

30 The Scottish Office Land Reform Policy Group, ibid, para 2.7

31 See Hunter, James, 1997, who goes into considerable detail on how this might be achieved.

32 See *Scotland's Parliament* - the Devolution White Paper (Scottish Office, 1997).

10 Sustainable Scotland: The Energy Dimension

TONY GLOYNE and ALAN HUTTON

Introduction

Energy is ubiquitous, it permeates every facet of society and respects no administrative boundaries. Accordingly it is almost impossible to discuss it out of context. It is a factor in planning, transportation, environment, health and welfare. Within the context of 'sustainability' it represents a vital dimension, involving as it does the basic thermodynamics of existence and progress. In the long term the present patterns of energy supply and use in Scotland are not sustainable. In the shorter term moves towards sustainability are possible. So far the development of such strategies has been limited, in the context of the Agenda 21 debate, to the issue of global warming and the anthropogenic generation of greenhouse gasses, with specific targets set for the reduction of CO_2 emissions. In this chapter we review the case for a more concerted sustainable energy strategy and comment on the problems and prospects for such a policy development in Scotland.

We begin with a brief consideration of the concept of sustainable development, its possible meaning at a national level and the general implications which follow for environmentally-oriented policy in the energy sector. In a review of present and future supply and use of energy in Scotland, we highlight the concerns which arise in relation to environmental sustainability, and summarise how the present energy policy regime impacts on these concerns. After noting some important problems for current UK and Scottish policy in the interactions between the realms of energy and environmental policy, we consider the developing policy context of a devolved Scotland as a framework for a more sustainable energy policy, indicating the likely limitations on how far this agenda can be pursued at a Scottish level. We end with some specific suggestions for possible elements of a more sustainable energy policy for Scotland, and conclude that although there are inherent limits to the effectiveness of policy at a national or regional level, devolution could offer opportunities for more environmentally effective energy policies.

Sustainable Development and Energy: General Issues

From the time of Malthus people have engaged in speculation about the long term prospects for the planet. Predictions have tended to be pessimistic yet, so far, the track record of the 'eco–doomsters' has not been impressive. Are current concerns further examples of crying wolf or for real this time? At a global level, two major challenges face societies: the relief of poverty and the maintenance of the ecological health of the planet. The conventional view has been that poverty is relieved through following a path of economic growth. That involves industrialisation which, in turn, is fueled by energy consumption. The crux of the difficulty is that whilst the very process of economic growth may threaten the environment, to curtail growth has been deemed wholly unacceptable in both developed and underdeveloped societies. In the face of this dilemma the Bruntland Report (1987) developed the concept of 'sustainable development', substituting 'growth' with 'development', with the caveat that development should be compatible with environmental and resource constraints so that the concept of environmental equity over time can be satisfied. Although sustainable development gained widespread acceptance at a political level, there remains no agreed operational definition and thus no specific agreement on what its policy implications might be. Instead we see a range of proposed measures aimed at 'buying time' by reducing the extent to which current patterns are unsustainable, involving an emphasis on 'no regret' measures.

Some analysts (including Beckerman, 1995) dispute the need for any adjustment, arguing that proponents of the precautionary principle base their case on thin evidence and fail to take account of technological ingenuity in meeting the challenge. A recent report (New Scientist, 1998) presents encouraging results from development work on catalytic solar-powered and water splitting techniques which could herald the hydrogen economy, for long a favoured system. Advocates of the technical fix could also cite nuclear fusion as a coming technology but, as yet, our extraterrestrial fusion reactor (the sun) may be a more prudent bet. This was certainly the view eloquently expressed over 20 years ago by Amory Lovins (1977) who counselled against going down the hard (nuclear power) road. Even amongst proponents of sustainable development there are important divisions of opinion on the adoption of 'strong' or 'weak' versions of sustainability.

Irrespective of such arguments some facts are not disputed, most importantly that the planet approximates to the thermodynamic definition of a closed system. The implications of this disturbed the 19th century thermodynamicists but the message only seemed to re–enter the economists' world with Boulding's famous 1966 paper, introducing the image of 'spaceship earth', and with the work of Geogescu Roegen (1971). Their arguments were given a higher, but more controversial, profile by the work of Meadows and his colleagues. *The Limits to Growth* (1972) and its successor *Beyond the Limits* (1992) carried a message broadly similar to sustainable development but the focus of their proposed solutions was on redistribution rather than the politically more acceptable combination of growth with redistribution.

The key energy issue is that developed economies are predominantly reliant on massive inputs of *finite* fossil and fissile resources. Thus current energy policies are in the long run inherently unsustainable. Critically, the vital resources fall into the non–renewable, non–recyclable category: once 'burnt' their potential is irretrievably lost. It may turn out to have been fortuitous that the CO_2 generated in fossil fuel consumption has, through the concern over global warming, led to a political climate in which international agreement on abatement targets may provide the breathing space for a fundamental reconsideration of energy policy. In the absence of any clear long run alternative, there appears to be little choice but to begin a transition from an over-reliance on fossil and fissile energy towards a policy based on renewable energy sources. As non–renewables are depleted such a strategy would combine the build-up of a stock of renewable energy capital, with major efforts to manage demand and to develop the 'fifth fuel'- energy conservation. In effect this means a shift away from a demand–led approach to one conditioned by supply–side constraints.

Sustainable Development and National Boundaries

Despite a lack of clarity as to its exact implications, it is clear that sustainable development includes both positive and normative dimensions. Two main arguments arise: the ethical proposition that provision should be made for a yet-unborn generation, and the efficiency argument that in some specific sustainable development sense, aggregate social welfare is maximised over time.

A key issue is that of stewardship over resources and the legacy to future generations. It is of course true that in any country where net capital stock per capita is increasing, that legacy is a positive one: a situation which is, *prima facie*, consistent with sustainable development. A counter argument suggests that concepts of capital are flawed and that conventional economic measures of national income do not adequately reflect the depreciation of environmental capital.

The basic thrust of the environmental lobby's case is that the entire framework is wrong. Proponents of 'weak sustainability', and probably most economists, are willing to concede that as an indicator of quality of life, let alone sustainability, national output measured as GDP per head is misleading. 'Weak sustainability' involves the assumption, made by Atkinson and Pearce (1993) in constructing their sustainability indicator, that manufactured and environmental capital can be substituted for each other: depreciation of the one can be compensated for by a net increase in the other. Quite how to value the depreciation of environmental capital remains unclear especially since some of that depreciation may cross national boundaries. Advocates of 'strong sustainability' dispute the notion of substituting manufactured for environmental capital and in consequence take a much tougher line on the maintenance of environmental capital.

Two questions emerge. First, is sustainability essentially a system or a sub-system concept? And, second, since there is no rate at which non-renewable resources may be depleted without 'depreciation' taking place, is 'strong sustainability' a position that can be taken seriously? What then does 'sustainable Scotland' mean? And can any economy where predominantly fossil and fissile sources are used to augment 'ambient' energy be deemed sustainable?

As a system planet earth has to be self-sufficient (solar energy excepted), but at a sub-system level national self-sufficiency has been abandoned in favour of securing the gains from specialisation and trade. Self–sufficiency may be possible for the UK as a whole, but harder to achieve in a small nation. Clearly if all countries are both self-sufficient and sustainable then the entire system would be sustainable. But is self-sufficiency a necessary condition for sustainability? Is 'sustainable trade' the more relevant concept? Conventionally, balance of trade equates cross-border value flows, but sustainable development might imply additional materials balance flows. The resulting depletion of non–renewable energy sources suggests that such a balance would not be sustainable. In that context should Scotland be a net importer or net exporter of energy? What interpretation should be placed on the Agenda 21 sentiment in favour of greater self–sufficiency at the local level? If transport, an energy intensive activity, shows steep cost increases then that would become a significant barrier to trade and encouragement to the pursuit of self–sufficiency.

Energy Demand

While energy is an input into all processes the exact nature of the links between energy consumption and quality of life is questionable. In the past it was assumed that any threat to growth and energy supply was tantamount to a direct attack on standards of living. British policy was therefore driven by forecasts of economic growth. However, there has been a progressive reduction in the energy ratio (primary energy demand/GDP) to the point that by 1995 it had fallen to less than two-thirds of its 1970 value In the UK. It is forecast to fall below half the 1970 value by 2020 (Department of Energy, 1995). Whilst this might seem encouraging, a proper interpretation of the trend raises complications since, over the period, there have been major changes in both the fuel mix, notably the dramatic increase in the use of natural gas, and in the composition of national outputs. The result is that the energy ratio value overstates the improvement. Still more sobering, however, is the fact that each scenario set out in Energy Paper 65 (Department of Energy, 1995) predicts a rise in UK CO_2 emissions from 1990 levels, including the low economic growth/high energy price scenario. A further feature worth noting is that a faster rate of economic growth, even with low energy prices, generates a greater fall in the energy ratio because of a more rapid turnover of capital stock. Although these forecasts pre-date Kyoto, little in the way of radical change is yet evident.

There is, however, growing doubt about the second link, that between material GDP and quality of life. The very use of 'development' rather than growth in sustainable development recognises the problem. This is nothing new to economists, as the work of Mishan (1967), Scitovsky (1976) and Hirsch (1977) indicates. A useful summary may be found in Anderson (1991). More recently the specific problem of the failure under standard national accounting conventions to include any allowance for the depreciation of environmental capital (including stocks of fossil, fissile fuels) has led to an upsurge in such work. Interestingly, the outcomes of recent efforts to compile an 'Index of Sustainable Economic Welfare' (ISEW) for Scotland suggest that growing GDP is coincident with a falling ISEW (Moffat & Wilson, 1994), although it might of course be true that lower GDP growth would have resulted in even lower ISEW.

Energy Supply in Scotland: the Current Position

Given the topical concern with energy matters it is surprising that specifically Scottish energy statistics are elusive, although 'in house' data exists in the Scottish Office. The Digest of UK Energy Statistics (DTI, 1997a) together with the complementary 'Brown Book' (DTI, 1997b) which covers the UK Continental Shelf (UKCS) oil and gas, provide comprehensive coverage.

Energy demand is split into four basic sectors - domestic, industrial, transport, and commercial - but since the ultimate objective of an economy is to improve quality of life to households, the proper distinction should be between direct and indirect domestic demand. In relation to sustainable development it is not the present but the long–term that matters most, and the issue is how to move from the current reliance on diverse fossil and fissile fuels to an economy in which renewable energy sources combined with extensive recycling radically reduce the dissipative nature of economic activity.

By any normal standards Scotland, with its claims on a major share of UKCS oil and gas as well as extensive coal deposits, is an energy-rich country. With the 'dash for gas' as the UK's favoured instrument in the search for CO_2 abatement coal production has withered. Along with the economic and social costs, that shift has involved a loss of coal resources. The switch to cheaper open cast mining has similar effects, together with other environmental and planning implications. The growing demand for oil and gas is now a concern.

Estimates of time to depletion have been consistently wrong and the economic theory of depletion provides limited guidance. For what they are worth, at present UK rates of exhaustion and current estimates of available reserves, oil will last between 10 and 30 years and gas for 20 to 40 years. The tendency however has been for new discoveries to keep pace with consumption, although the average size of new finds is falling. Low oil prices ($14/barrel in March 1998) give little incentive for UKCS exploration and, as technologies continue to develop, the above estimates must be treated with caution.

Scotland enjoys an overcapacity in electricity generation in the region of 100 per cent relative to simultaneous maximum demand. Roughly half of all electricity comes from the nuclear stations at Hunterston and Torness conveniently located to serve the major conurbations of Glasgow and Edinburgh. Some 10 per cent of electricity is currently derived from renewable resources, predominantly hydro. Scotland is a significant net exporter of electricity through the inter–connector to England and with further prospects once connection to Northern Ireland is established. In terms of CO_2 abatement nuclear power has obvious merit and is a reliable source of base load. With the proven breeder cycle it represents a potential energy resource comparable in scale to coal. Unfortunately it is not a renewable resource and the accompanying problem of nuclear waste management makes nuclear power environmentally controversial. Nevertheless, it may yet have an important future role to play in the transition towards a renewable energy base.

Our broad conclusion might be that in the medium term Scotland has no obvious problem of energy supply. That is, however, no reason for complacency since sustainable development poses the potentially huge problem of making the transition to an energy base with a very different composition in a short time scale when measured against lead times for the adjustment of capital stock.

With 80 per cent of the population living in urban areas energy supply has taken the form of delivering 'concentrated' energy to major load centres. Given the extensiveness of renewable energy sources the long–run viability of these urban centres may even come under question. Arguments based on agglomeration economies may no longer apply with the same force. Current patterns of economic activity reflect the technological requirements associated with non–renewable energy sources. The long–term challenge of sustainable development is the efficient transition from a fossil/fissile based, demand-driven energy supply to a supply-constrained pattern of energy demand (and of economic activity) reliant upon renewable sources.

Scotland's Renewable Energy Potential

With an exposed west coast offering wind and water power in abundance and a relatively low population density affording scope for bio-fuels, Scotland's renewable energy potential seems high: it could even be enhanced by the likely climatic effects of global warming. Critical to realising that potential, however, are questions of cost, including the point at which, once prime locations have been exploited, less favourable sites will bring increasing marginal cost.

Renewable energy comes in various forms ranging from the use of passive solar techniques for space heating through to specific power generating techniques from direct and indirect use of solar and gravitational energy. It has two characteristics of profound significance for policy. Firstly, whilst the energy has a high intrinsic quality it is relatively dilute and therefore has to be gathered on an extensive basis. Secondly, the flow of renewable energy is uneven, wind power for example being proportional to velocity cubed. Accommodating these major fluctuations in supply is difficult, particularly because of the problems of storage of electrical energy. This points to a further question, namely, would the aim be to replicate existing energy vectors from a renewable base or will the pattern of end use be required to adjust to the nature of supply? At present Scotland consumes a range of fuels, each with its own characteristics. To the motorist anthracite is not a substitute for petrol - in other words, energy equivalence should not be equated with substitutability. Inter-fuel conversion may become a very significant issue given the high investment in energy using infrastructure and devices and the fact that a high proportion of identified renewable resources take the form of electricity.

Scotland's energy potential cannot be estimated in any precise way. It depends, amongst other factors, on price. Although high energy prices do encourage some conservation and depress demand, the currently available evidence suggests that price elasticities of demand (an economist's measure of responsiveness to price changes) are low. Very large price rises would, therefore, be needed to induce even modest reductions in energy demand, and, as was highlighted in the recent debate over VAT on domestic fuel, such price increases are highly regressive in their impact on households.

While there is no dispute over the existence of significant potential sources of renewable energy in Scotland, the economic potential is rather less clear cut. At present wind power is the favoured source but even here a theoretically huge resource is reduced to a much more modest contribution after planning and other economic considerations are taken into account. This certainly was the conclusion of a study jointly undertaken by the Scottish power companies, the Enterprise Boards in Scotland, the DTI and the Scottish Office (Scottish Hydro-Electric et al, 1993). Their report suggests that wind power could provide roughly 50 per cent of Scotland's current electricity supply (see Table 1). At that time wave power was not deemed a proven technology on a significant scale, a conclusion which must surely be revised at a later date. However, the study was primarily concerned with integrating renewable electricity into the grid system and did not explicitly address the broader issue of the potential for sustainable energy.

Table 10.1 Potential Electricity Supply from Renewable Sources in Scotland
Source: Scottish Hydro-Electric et al (1993), p.21

Maturity of technology	Technology	Lifetime	Theoretical resource		After environmental, planning and practical considerations	
		years	MW	GWh/y	MW	GWh/y
Near to actually operationally proven	New Hydro	50	1,000	3,400	400	1,400
	Waste combustion	20	400	3,200	70	500
	Sewage gas	30	10	90	0.2	3
	Landfill gas	15	10	90	10	90
	Wind	20	100,000	266,000	7,300	19,000
Near operationally proven to demonstration	Farm wastes	20	30	700	90	700
	Energy crops and forestry	20	2,300	17,000	200	1,600
Demonstration to research	Wave (mainly inshore)	30	590	1,300	300	691

We endorse many of the points made in the more optimistic report *Renewable Energy: Scotland's Future* (Friends of the Earth, 1992). Whilst in the short–term we do not anticipate a rapid switch away from cheap and abundant fossil/fissile sources, a Renewable Energy Development Agency would help advance such a switch, in the long-run. Increased investment can bring the added benefit of a stimulus to growth in Scotland in the relatively new environmental technologies sector.

In the end sustainable development only works for the world economy as a whole. Decisions which determine the levels and patterns of energy generation, supply and use in Scotland should contribute towards that overall goal. The ways in which energy–related activities impinge on the natural environment, threatening sustainability, are significant and can be classified into:

- the depletion of non–renewable resources
- CO_2 emissions effecting world climate change

- other atmospheric and water pollution impacts
- 'fuelling' processes leading to bio–diversity loss and environmental degradation.

As we have noted, the recent pre–occupation of policy in Britain and around the world is on the second of these impacts.

Sustainable Energy Policy in Scotland: The Policy Context

Objectives of a sustainable energy policy

Any effective strategy for sustainable development will consider together the full range of economic, social and environmental effects, both long-term and short-term; and will recognise path dependency: the irreversibility of many effects. Since the use of energy underlies the provision of the whole range of goods and services aimed to further economic and social improvement, energy demand is influenced both by the demand for these final goods and by the efficiency with which energy is used in their production and consumption. An environmentally sustainable energy policy might, most obviously, involve:

- reduction of end user demand for energy;
- more efficient generation of energy from non–renewable sources;
- increase in use of renewable primary energy sources.

Current trends in energy use are, in general, moving in the opposite direction. Barring any unforeseen technical developments (with the possible exception of fusion) there would seem, in the long–term, to be no alternative to the increased use of renewable sources. Consideration might by that stage be given to the movement of population to the energy sources, and the substitution of localised for networked supply.

At any time the level and pattern of energy use are the product of decisions made by households, businesses, public agencies, and central and local government departments. Choices and constraints interact unpredictably and public policies in a number of fields can have unintended consequences for energy use. In the UK - and for Scotland - the achievement of a high degree of complementarity between energy related policies by the DTI, the DETR, the Treasury, and the Scottish Office has proved difficult. At the moment such policies remain relatively uncoordinated and, in important respects, contradictory. Conflicts arise between goals in the fields of national economic competitiveness and social policy, which have led to attempts to reduce unit energy costs, and environmental sustainability which requires conservation and improved energy efficiency and may point to higher energy prices.

In general, regulatory and tax policies (with the exception of taxes on petroleum products) have been geared to the reduction of energy prices to the consumer. Policies are essentially 'market conforming', founded on the belief that (with some limited adjustments) competitive market mechanisms will achieve the social good. Thus for example in the 'dash for gas' in electricity generation, energy efficient power generation is being pursued through a 'no regrets' policy which makes short-term economic sense whilst having negative environmental spin-offs particularly on CO_2 emissions. These advantages may distract attention from the longer term problem of a greater underlying rate of depletion of fossil fuel sources. Notwithstanding the achievement of CO_2 abatement, total UK energy consumption reached a new high in 1996. Thus, for the UK as a whole, there is a lack of strategic policy in both energy and environmental fields.

It is, of course, difficult to confine environmental policy relating to energy to the level of a small nation such as Scotland. This follows from the nature of energy sources and uses discussed above: external effects from energy use do not respect national boundaries, whilst an energy strategy heavily influenced by the goal of improving national economic competitiveness through cost reduction will generate a collective neglect of external environmental costs. Scotland has no choice but to work within the constraints of economic and environmental policies at the UK and European levels. As an open trading economy it is also clearly exposed to the impacts of economic and environmental changes in the wider world and to global market pressures which may tend towards lowest common denominator outcomes in environmental areas. Policies for a sustainable Scotland cannot be sustained purely at the Scottish level. Goals, targets, and regulation at the EU level and wider are already significant, and tax powers as well as the framework of energy utility regulation fall, at the moment, within the powers of the UK government.

Many energy-related elements of policy, including housing, planning, transport, and environmental protection, are already within the remit of the Scottish Office and here, as in other areas, the Scottish Parliament and Executive may offer the opportunity for radical change. Scotland's small scale may prove advantageous as it already has a relatively tight–knit economic and political community involving frequent contact between elected and official representatives of central and local government, the leaders of industry, financial institutions, trade unions, educational and training bodies, and the churches. If the Scottish Parliament does provide the basis for a more consensual and participative style of politics, which could bring greater stability to Scottish government as well as rendering policy-making more accountable, then the possibility arises for achieving more coherent policies for sustainability in the long-term (see proposals in John Wheatley Centre, 1997).

Energy Policy and Sustainable Development in Scotland: Issues and Proposals

Much will depend on the way the Scottish Parliament and Executive operate, but the potential exists for an Environmental Policy Committee of the Parliament drawing on an expert advisory commission (a successor to the Secretary of State's Advisory Group on Sustainable Development) and a wider citizens' forum, to play a leading role in establishing a sustainable development dimension for policy across the board, including energy policy.

For the reasons already outlined, 'subsidiarity' is a sensible strategy in this field. Action Programmes at the EU level will become increasingly important. The implementation of EU directives would be overseen by government at the Scottish level and incorporated into Scottish energy/environmental policy. A fundamental shift in the base of taxation away from income and towards energy and resource use is desirable. But given all the assumed dangers of social and environmental 'dumping', it is probably not appropriate for green taxation powers in general to be exercised only at the Scottish level. Nevertheless Scotland will have a wide range of powers of regulation, investment, subsidy and, in some ways most important, national direction-setting and education. A pre-requisite for any satisfactory consideration of a sustainable energy policy is the publication of a full set of energy statistics for Scotland.

Amongst Scottish policy areas with significant impacts on energy use, transport and land-use planning are explored elsewhere in this volume. Here, we focus on issues concerning housing, energy efficiency, fuel poverty and regulation of the energy utilities.

Housing, energy efficiency and fuel poverty

The problem of fuel poverty - the inability of people to afford adequate heat in the home - was brought into sharp focus by the extension of VAT to domestic fuel. The nature and extent of fuel poverty in Scotland and its implications for the quality of peoples' lives have been extensively documented by Energy Action Scotland (1997). The problem results from the interaction between income, housing quality, and energy prices. The longer term solution must rest with a combination of increases in disposable income and huge improvements in the energy efficiency of the housing stock. On the National Home Energy Rating (NHER) scale of 1 to 10, 17 per cent of Scottish homes score 2 or less and the 1996 average score was only 4.1. According to the 1996 Scottish House Condition Survey, more than 93 per cent of Scottish homes failed to reach the accepted standard for a modern home, a rating of 7 (EAS, 1997). Low income households tend to live in housing with the lowest energy efficiency and the highest incidence of dampness and related problems.

Reducing energy use by increasing energy efficiency is usually a capital intensive strategy. 'Retrofitting' is much more costly than incorporating energy saving in new construction, but a big increase in provision for existing housing stock within the Home Energy Efficiency Scheme is necessary (EAS, 1997). After a period of low levels of house building, especially in the social rented sectors, Scotland is left with a housing shortage along with the problem of the quality of the existing stock (SEF, 1997). A significant long–term housing investment programme is needed, and a concerted national strategy may be more possible following devolution. New investment offers the opportunity to build in radically improved energy efficiency, and an upgrading of Scottish Building Standards Regulations on insulation to Swedish or Danish levels would provide a 20 per cent-60 per cent improvement in energy efficiency. A typical Swedish house is 5° C warmer whilst using 25 per cent less energy than in Scotland (EAS, 1997).

The future of energy utility regulation in Scotland

The reduction of end–user unit energy prices is inimical to sustainability, yet that has been the main goal of the regulators in the privatised gas and electricity industries. Indeed a number of aspects of the structure and regulation of these industries reflect, at the very least, a lack of concern for environmental implications. The essential presumptions underlying UK energy regulation have been that, ultimately, free markets are better than publicly-owned regimes; that competition can be satisfactorily introduced and regulation should therefore be temporary; that competition will deliver benefits to all consumers through 'choice' and price reductions which are the final measure of benefit; and that there are few desirable public service goals which would not be achieved more effectively through the operation of competitive markets. Much controversy has surrounded the practice of regulation in the energy industries focusing *inter alia* on high profits and the Windfall Levy, 'fat cat' renumeration, and the regressive effects of some company policies. Regulation has even been seen as weak in relation to the goal of reducing prices.

Both OFFER and OFGAS have a legal duty to take into account the effects of activities they regulate on the environment. Within the electricity sector, an annual allowance equivalent to £1 per customer is allocated for use by the public electricity supply companies (PES) for approved energy efficiency schemes and, under the Scottish Renewables Obligation (SRO), funding is provided for electricity generation schemes using renewable sources. Whilst the third order under the SRO has now been issued, few of these schemes have yet reached the stage of significant generation. In the process of market liberalisation which culminates with the full opening of the domestic supply markets in gas and electricity in 1998, concern for environmental impacts 'has been an afterthought' (Eyre, 1996).

In the Scotland Bill utility regulation is a matter reserved for UK government. Given the importance of the energy sector, that represents an impediment to the development of an integrated sustainable development strategy for Scotland. There appears to be nothing in principle that should prevent the devolution of regulatory responsibilities to Scotland. Indeed the electricity industry in Scotland was privatised under separate legislation and retains a basically vertically integrated structure with the two PESs (Scottish Power and Scottish Hydro-Electric) retaining significant responsibilities for generation, distribution and supply. As Eyre (1996) points out a vertically integrated utility can have an overview of the full set of environmental implications associated with all stages of generating and supplying electricity and these can be taken into account in overall project planning. OFFER has a Scottish office which handles aspects of regulation relating to the different arrangements in Scotland.

Since in the immediate future regulatory responsibilities for the energy supply sector in Scotland are to stay with the DTI, it is vital to enhance the Scottish dimension in regulatory practice. The DTI review of the regulation process is likely to propose enhanced political accountability of the regulatory machinery if not the full integration of gas and electricity regulation. Such an approach to regulation of the energy utilities should improve the prospects for more sustainable environmental policies. In order to pursue the energy dimension of an integrated sustainable development strategy, it is vital that there is an Office of the Energy Regulation Agency in Scotland, that the consumer representative element in regulation operates at a Scottish level, and that the Agency is accountable to the Scottish Executive and to a relevant committee of the Scottish Parliament. The existing integrated structure of the electricity industry in Scotland should not be further eroded.

Whatever the extent of devolution of responsibility for energy regulation there is a strong case - not only for environmental policy reasons - for fundamental changes in the regulatory regime. We take the view that the presumptions of current regulatory practice listed above are inadequate or inappropriate. We follow Souter (1994) in arguing for a 'stakeholder' approach to regulation of the energy utilities, the relevant stakeholders being consumers, shareholders, managers, employees, new market entrants, suppliers, and government representing a national interest. Beyond a concern with the reconciliation of conflicting interests amongst the other stakeholders 'national' concerns will include social, economic and quality of life goals, the last including longer–term environmental sustainability.

From 1998 onwards, competition will extend to the domestic market in both gas and electricity supply. The full environmental implications of the change are unclear. Fuller competition may create pressures for more efficient resource use in generation and distribution, including greater use of renewable sources. On the other hand competition based on price-cutting runs counter to increased energy efficiency. Liberalisation also undermines aspects of the stable investment framework needed for the development of sustainable energy supply.

From an environmental standpoint the regulatory regime should not be weakened as competition increases. Promotion of energy efficiency should be a condition of all new supply licences. The SRO should be strengthened and what Eyre (1996) calls a Sustainable Energy Fund within the price levy should be introduced to fund large scale programmes for the management of energy demand. The implementation of the new EU directive on integrated resource planning, allowing environmental factors to be taken into account in investment decisions on electricity distribution, would also encourage a reorientation of regulation towards a framework in which companies see themselves as 'suppliers of energy services', with the goal of reducing consumers' energy costs more by energy efficiency measures than through the reduction of unit prices. The results of market liberalisation are likely to make sustainable energy strategies more difficult. A more proactive interpretation of the regulators' duty to promote efficient energy use on is vital.

Conclusion

Sustainable development, however defined, is a long term goal. Energy sustainability requires a progressive shift towards renewable energy resources. Scotland, a net exporter of energy, is energy rich both in fossil fuel and renewable energy sources.

Political devolution provides an opportunity to develop an integrated Scottish approach to sustainable development. Such an approach is desirable and could be practically effective. Energy policy should be a fundamental part of any sustainability strategy. At the Scottish level, however, the possibilities of influence over energy demand and supply outcomes are limited. Since leaving energy outcomes to market forces results in a neglect of environmental considerations, public policy initiatives are necessary. Some, particularly relating to taxation and international market regulation and standard setting, will inevitably be determined at UK, EU or higher levels. But important policy choices affecting energy efficiency and energy demand, as well as the development of renewable energy sources, do remain to be made at the Scottish level. In that context we have argued that it is vital for the implementation of a sustainable energy strategy that there is both a change in the approach to regulation of the energy utilities and a significant enhancement in the accountability of energy regulation to Parliament and Executive in Scotland.

In bridging the gap between present patterns of energy use and future patterns based on renewable energy sources, Scotland is fortunate in the availability of nuclear capacity and a significant supply of fossil fuels. In this chapter we have suggested a number of directions in which short and medium term policy in Scotland can move in order to aid that long run transition through promoting sensible demand side management and the technical and economic development of renewable energy sources whilst, at the same time, enhancing the quality of life of all citizens in the process.

References

Anderson, V (1991) *Alternative Economic Indicators,* Routledge, London.
Atkinson D & Pearce DW (1993) 'Measuring sustainable development', *The Globe* 13 June, UK GER Office, Swindon.
Beckerman, W (1995) *Small is stupid : blowing the whistle on the Greens,* Duckworth, London.
Boulding K (1966) 'On the economics of spaceship earth' in Jarrett H (ed) *Quality in a Growing Economy*, Resources for the Future/Johns Hopkins University Press, Baltimore.
Department of Energy (1995) *Energy Paper 65,* HMSO, London.
Department of Trade and Industry (1997a) *Digest of UK Energy Statistics* HMSO, London.
Department of Trade and Industry (1997b) *Energy Report* HMSO, London.
Energy Action Scotland (1997) *Scotland's Fuel Poor: Fuel poverty in a changing climate,* EAS, Glasgow.
Eyre N (1996) 'Meeting environmental objectives in liberalised energy markets' in Corry D, Hewett C & Tindale S (eds) *Energy 98: Competing for Power,* IPPR, London.
Friends of the Earth (1992) *Renewable Energy: Scotland's Future,* FoE, Edinburgh
Geogescu–Roegen N (1971) *The Entropy Law and The Economic Process,* Harvard University Press, Cambridge MA.
Hirsch F. (1977) *The Social Limits to Growth,* Routledge & Kegan Paul, London.
John Wheatley Centre, Commission on Environmental Policies (1997) *Working for Sustainability: an Environmental Agenda for a Scottish Parliament,* John Wheatley Centre, Edinburgh.
Lovins, A (1977) *Soft Energy Paths,* Penguin, Harmondsworth.
Meadows D et al (1972) *The Limits to Growth,* Universe Books, New York.
Meadows D et al (1992) *Beyond the Limits,* Earthscan, London.
Mishan E (1967) *The Costs of Economic Growth,* Staples Press, London.
Moffat I and Wilson MD (1994) 'An Index of Sustainable Development for Scotland', *International Journal of Sustainable Development and World Ecology,*
New Scientist (1998) 'Catalysts for change' Feb 28th.
Scitovsky T (1976) *The Joyless Economy,* Oxford University Press, New York.
Scottish Hydro–Electric et al (1993) *An Assessment of the Potential Renewable Energy Resource in Scotland,* Scottish Hydro-Electric.
Scottish Environmental Forum (1997) *Poverty and Sustainable development in Scotland,* Edinburgh.
Souter, D (1994) 'A stakeholder approach to regulation', in Corry, D, Souter, D & Waterson, M (eds) *Regulating Our Utilities*, IPPR, London.

11 Environmental Management: A Business Perspective

ALISTAIR DALZIEL

Introduction

For a business to project a positive environmental image, the development of a system of beliefs and values, rather than solely meeting legislative requirements is essential. Although compliance with legislation is assumed, this is unlikely to be a sufficient basis on which to adhere such a profile. Companies wish to project such an image for a number of reasons. These include:

- Enabling future expansion
- Employee motivation and loyalty
- Better recruitment prospects and retention of the right kind of people
- Improved public reputation
- Increased customer loyalty

Many organisations wishing to project the right environmental image will have a strong regional or local presence. For them, the opinion of the community is particularly relevant whether during the process of obtaining planning consent for new premises or expanding existing plant, and for the maintenance of high class recruitment. There is no business activity which cannot improve its environmental performance and examples of good practice, where businesses attempt to demonstrate their environmental credentials include:

- Minimisation of wastes
- Appropriate use of recycled products
- Energy conservation
- Car fleet management
- Purchasing/supplier policies
- Environmental auditing

Failure to address these issues will work to the disadvantage of financial performance through, for example, costs arising from non-compliance; loss of profile and thus market share; or indeed through last operational cost saving opportunities.

Voluntary disclosure of environmental information is becoming increasingly popular, particularly among larger organisations. For reasons such as budgetary constraints and restrictions on management time, small and medium size companies may find it difficult to address environmental issues with the same vigour (see also Chapter 6). However, there are elements of best practice which smaller companies can emulate. There are also good reasons for developing a proper understanding of the environmental impact of any business and making that information publicly available, even where it is not a legal requirement.

The advantages include not only financial gains through, for example energy savings, reduction in raw material consumption and minimisation of wastes generated, but also through enhancing the company's credibility in dealing with all its interest groups whether customers, shareholders, employees or environmental activist groups. Information provided directly from the industry also encourages a better informed public. Voluntary dissemination usually allows a company to present information in its own way, to audiences for whom it is relevant, in ways which are timely and in forms which make it comprehensible. The main elements of best environmental practice are:

- Providing a relevant policy
- Setting easy to implement procedures
- Setting programmes with clear objectives

The environmental policy must be concrete and identify measurable goals rather than simply producing words of good intent. It must relate to the contribution of good environmental practice to meeting core business or organisational goals. The policy must also be turned into action through the development of procedures and programmes. Regular environmental audits will allow close monitoring of the performance of such programmes.

The term 'environmental audit' is increasingly used to refer to a systematic examination of an individual site carried out within an overall company wide audit. The audit is designed to provide management with information on a regular basis on whether the company's policies, procedures and programmes are being achieved and the extent to which the company is in a position to comply with existing and developing environmental legislation. In other words, the results of an environmental audit provide the information that allows managers to measure the overall environmental performance of the company. The different types of audit can be categorised as follows:

- **Technical review**

Such an audit will involve the systematic collection of information about the existing and potential impact of the company's activities on the environment. It will normally cover compliance with pollution control and waste management legislation. It will not cover management practices.

- **Management review**

This will focus more on management procedures and record-keeping and will also gather information on compliance with legislation. It may also review procedures in the context of company policies, programmes and other requirements. It will not examine the existing or likely impact of the operation on the surrounding environment from a technical standpoint.

- **Due Diligence review**

This will examine the likely cost of implementing pollution control and site remediation actions and will take account of existing and future legislation. Such liability reviews are normally carried out in the context of mergers, acquisitions and long term company planning.

Due Diligence investigations are routinely undertaken when commercial transactions are being considered. In the past these investigations have focused on markets, trading, tax, pensions and legal considerations. Because of increasingly stringent regulation, however, Environmental Due Diligence now forms an important component of the investigations.

Environmental liabilities are unlike any other type of liability facing a company. First, it is often open to interpretation whether a company is compliant with environmental legislation. Second, when a company is non-compliant, the risk it faces in terms of financial exposure is difficult to quantify and the penalties and costs of remedial action can be extreme. Consequently, financial institutions now require environmental Due Diligence to be conducted on a much more frequent basis than in the past. The purpose of the assessment is to identify the environmental issues that are material to the proposed transaction. Environmental Due Diligence comprises the following:

- Identification of material liabilities
- Assessment of legal compliance

- Evaluation of necessary remedial action
- Estimation of associated capital or revenue costs

Although such audits are commonly undertaken on large transactions involving heavy industry, they are now beginning to be undertaken on smaller transactions affecting property as well as manufacturing and service industries. The message is do not be complacent: most businesses have the potential to have a harmful effect on the environment.

When considering a transaction, many companies initially will commission a Screening Study. Typically a detailed environmental Due Diligence will be commissioned if either the past or present use of a facility could give rise to significant liabilities. This is especially critical if the operation or property is situated in an environmentally sensitive location. When environmental liabilities are identified, mechanisms should be evaluated to minimise risk and allow the transaction to proceed. Questions that need to be answered include:

- Should that part of the business holding the liability be excluded from the transactions?
- Would the liability become more acceptable if the price to be paid for the business was adjusted to allow the speedy implementation of remedial measures?
- Are the Vendor and Purchaser able to agree a package of indemnities and warranties?

Once the equity in a company has been acquired, a financial institution will be looking forward to an exit at some time in the future. Getting it wrong and failing to quantify the environmental liabilities at the time of completion carries the risk of jeopardising the exit strategy - particularly if the preferred exit is a flotation rather than trade sale. A properly implemented environmental audit plan can, therefore, provide a range of benefits for the organisation. These are to:

- Provide a framework for measuring and managing environmental performance
- Reinforce accountability for the environmental dimension of the business. The audit process requires managers to be clear about their responsibilities and how these are being implemented.

- Raise awareness of the importance of professional environmental management throughout the organisation.
- Provide valuable information for future planning. This includes the future design of processes and products as well as inputs into future financial plans where pollution control and other investments are to be made.
- Allow senior managers to feel secure that all the environmental aspects of their business are being professionally managed. It will not remove all possibility of environmental incidents but will at least reduce the likelihood and associated costs (since the extent of damage should be smaller).

Having identified the environmental issues, implementing a system of management in accordance with one of the regulated programmes (such as EMAS, BS7750 and ISO 14001 could be undertaken. See Chapter 6). Alternatively, a number of companies are content to establish their own in house systems as a stand-alone initiative.

Whatever a company's strategic objectives or practices, in order to remain competitive it is important to have an environmental policy which goes beyond a well intentioned statement in the Annual Report and simply complying with minimum regulations. Good environmental management usually means good business.

12 Business Strategy and the Environment: Implementing European Environmental Management Systems

PETER A. STRACHAN

Introduction

For many Scottish business organisations the implications of uncertain and complex environmental problems are driving the need for improved management approaches (Schmidheiny 1992). While 'command and control' legislation has traditionally been seen as the most effective way to improve environmental performance standards, this approach has recently been overtaken by a more managerial one. This focuses on the formal management systems of the organisation (Strachan and Moxen 1998; Bris and Kok 1995, 1994; Hillary 1995; Moxen and Strachan 1995). The recognition that the management systems of an organisation have an important role to play in the raising of environmental performance standards has led to the formulation of two new international environmental management systems (EMS):

- The International Organisation for Standardisation (ISO) 14001 EMS. This is available for voluntary participation by business, not-for-profit and public organisations around the world. ISO 14001 is a voluntary 'mega-standard' and represents for the first time an initial global benchmark for system-based EMS.
- The European Union (EU) Eco-Management and Audit Scheme (EMAS). This is available for voluntary participation by manufacturing organisations across the Member States. The EMAS represents a significant step forward in European environmental policy, law and management which parallels the international ISO 14001 EMS development.

These EMS are currently receiving much interest from governments around the globe, national and international industry led-bodies and a large number of organisations operating in a wide range of economic sectors. ISO 14001 and the EMAS Regulation have significant implications for the way in which Scottish companies are managed and organised. They commit those who participate to fundamentally rethink their core business activities in order to make them more consistent with the macro-level goals of 'sustainable development' (Gelber *et al* 1997; Riglar 1997; Strachan 1997a).

In the short time during which these EMS have been on the strategic and operational agendas of private, not-for-profit and public organisations, a substantial literature has been produced on the general merits of system-based EMS and on the contribution that they can make towards a more sustainable society (Hillary 1997; Spencer-Cook 1997; Strachan 1996; Biondi and Frey 1995; Counsell and King 1995; Franke and Watzold 1995; Institute of Environmental Management 1995; Faure *et al* 1993). Unfortunately, this growing core of knowledge provides little empirical evidence about the benefits and potential implementation issues which organisations may face when pursuing this route to ecological enlightenment (Strachan *et al* 1997b). The purpose of this chapter is to address this oversight in the existing business and environmental literature and to present the findings from one of the first surveys of enterprises to have implemented the EU EMAS Regulation in the UK. A discussion of the findings are presented once the argument is set in historical context. The chapter concludes with a number of recommendations which the Scottish Parliament may wish to consider in the effective promotion of business EMS.

The Rationale and Aims of the Study

The UK chosen for study because the British Government was a champion of the EMAS Regulation despite much resistance from other EU Member States (Franke and Watzold 1995). Furthermore, the UK was the first Member State to fully participate in the Regulation's implementation. For these reasons it was felt that important lessons may be learned about the implementation of EMS in UK organisations that could assist organisations in other member states to integrate the Regulation into their strategic decision-making and operational activities more fully (Strachan 1997b). Furthermore, the EMAS and ISO 14001 EMS share the same aims and despite some minor technical differences their main components are very similar. Therefore, it is anticipated that the study's findings are transferable to a range of organisations which are in the process of implementing ISO 14001 and that they will also be able to gather valuable lessons on the implementation of EMS. The aims of the study can be grouped into three clusters:

- *Defining the Main Triggers for Implementing the EMAS Regulation*. The first cluster examines the reasons why UK organisations implemented the EMAS Regulation. It considers whether the organisations have been induced to introduce the EMAS Regulation by stakeholder pressures outside the organisation such as legislative and regulatory pressures; and/or internal motives including organisational pressures to reduce current operating costs by seeking 'green' improvements in manufacturing processes.
- *Identifying the Main Changes Introduced in the Organisation's Management Systems*. The second cluster examines the implementation of the EMAS Regulation into the organisation's internal management systems. This included an investigation into the main managerial, organisational and technical issues which these organisations needed to work through in their attempt to implement the EMAS Regulation into their strategic and operational activities.
- *Providing An Analysis of Current and Future Benefits*. The final cluster examines the organisational and societal benefits that UK organisations claimed to have accrued through their participation in the EMAS Regulation. The study examined whether the EMAS Regulation allowed the organisations to increase market share, enhance their green image or improve their internal management.

Before going on to present the findings, the background to the formation of the EU and its policies on the environment are reviewed to set the chapter into context.

Developments in the Creation of the European Union

The period following the Second World War witnessed a number of attempts to stimulate integration among European states, including the negotiation of the European Convention on Human Rights and Fundamental Freedoms (1950) and plans to promote political, economic and military co-operation. Most of these attempts failed with the notable exception of the Treaty of Paris (1952), which established the European Coal and Steel Community (ECSC). Six European States participated in this organisation -

Belgium, France, Germany, Italy, Luxembourg and the Netherlands - which was designed to place the production of steel under joint control. The general success of the ECSC stimulated interest in a more comprehensive form of economic integration and two further agreements were signed in Rome (1957), namely the European Economic Community (EEC) Treaty and the European Atomic Energy (Euratom) Treaty which was intended to create a specialised market for atomic energy.

The EEC Treaty, on the other hand, was a rather more comprehensive agreement which aimed to promote economic integration between the original six member states. The 1957 Treaty of Rome stated that the task of the Community was to establish a common market, progressively approximating the economic policies of the Member States to promote throughout the community:

> ...a harmonious development of economic activities, a continuous and balanced expansion, an increase in stability, an accelerated raising of the standard of living and closer relations between the states belonging to it (Wolf and White 1995, p.25).

At this stage in the development of the EEC, environmental issues were not addressed and the term 'environment' did not appear in the formal documentation of the Treaty of Rome. This is not surprising as environmental problems were barely recognised in the 1950s by the war-damaged economies of Europe, with the need to fundamentally reorganise industries and to promote self-sufficiency in food supplies. During the course of the more affluent 1960s and 1970s it became apparent that the achievement of the Treaty of Rome's aims could no longer 'be imagined in the absence of an effective campaign to combat pollution ... or of an improvement in the quality of life and the protection of the environment' (Sunkin et al 1998, p.11). Furthermore, the 1972 Heads of State and Government meeting stated that:

> Economic expansion is not an end in itself. Its firm aim should be to enable disparities in living conditions to be reduced ... It should result in an improvement in the quality of life as well as in standards of living. As befits the genius of Europe, particular attention will be given to intangible values and to protecting the environment, so that progress may really be put at the service of mankind (Sunkin et al 1998, p.11).

During the last two decades the EEC has grown from the initial membership of six to fifteen. Following German reunification in 1989, East

Germany was absorbed in the Community, subject to a number of transitional arrangements. Not only has the community grown in size in recent years it has also considerably increased its spheres of influence by becoming much more than an association of states pursuing economic goals. Since 1993, the European Community has been superseded by an even more comprehensive European organisation which spans not only economic and political activities but also social, monetary, cultural, environmental, and health and safety affairs. Today the basic principles which underpin the EC's policy on the natural environment are summarised by Mathijsen (1995, p.347):

- It is a shared responsibility between the Community and the Member States; they are fully responsible for the implementation of the policy;
- It is an area where the principle of subsidiarity should apply;
- Since it generally involves trans-border problems, harmonisation is essential;
- Homogeneous standards throughout the Community are required to avoid barriers to trade and distortions of competition; and,
- The Community objective of improvement of the quality of life requires a high level of environmental protection.

Mathijsen goes on to state that the objectives of the Community policy on the environment are:

- Preservation, protection and improvement of the quality of the environment;
- Protection of human health;
- Prudent and rational use of natural resources;
- Promotion of measures at international level to deal with regional and world-wide environmental problems.

Following amendments to the Treaty of Rome by the 1986 Single European Act (which established for the first time that the Community could take legitimate action in the field of environmental protection) and the Treaty on European Union (1992) the general aims of the EU are more widely stated and more fully reflect the importance of the natural environment in EU Member States:

> The Community shall have as its task, by establishing a common market and an economic and monetary union and by implementing the common policies of activities referred to in Arts 3 and 3(a) to promote throughout the Community a harmonious, and balanced development of economic activities, sustainable development and non inflationary growth, respecting the environment, a high degree of convergence of economic performance, a high level of employment and of social protection, the raising of the standard of living and quality of life, and economic and social cohesion and solidarity among Member States.' (Wolf and White 1995, p.25).

Within this definition 'sustainable development' is defined as:

> a policy and strategy for continued economic and social development without detriment to the environment and the natural resources on the quality of which continued activity and further development depend' (Sunkin *et al* 1998, p.26).

More broadly, the definition of sustainable development provided by the Brundtland Report is quoted with approval within EU circles and this states, 'development that meets the needs of the present without compromising the ability of future generations to meet their own needs' (Sunkin *et al* 1998, p.25).

From 'Command and Control' to Market-Based Measures

As outlined above, before the Single European Act (1986) entered into force, there were no provisions in the Treaty of Rome concerning the environment. However, this did not prevent the Community from taking action in this field and the principles of the EU's environmental policy are set out in action programmes on the environment which date back to the early 1970s. The EC's first Action Programme on the Environment laid down the basis for what has become known as the principle of preventative action. This states, 'the best environmental policy consists in preventing the creation of pollution and nuisances at source, rather than subsequently trying to counteract their effects' (Sunkin *et al* 1998, p.31)

Following the launch of the EU's First Action Programme in 1973, Europe has witnessed a rapid growth not only in environmental awareness among the general public, government and industry, but also in the quantity and coverage of 'command and control' environmental legislation both at the national and EU level. Over the last two decades, the EU has established four subsequent action programmes on the natural environment and more than 200 pieces of environmental legislation covering air, water and land, as well as many products, processes and environmental impact assessments (Hillary 1995).

More recently, the Fifth Environmental Action Programme (EAP) running from 1993 to 2000 has reinforced the combined roles of companies in economic development and environmental protection (CEC 1992). This Fifth Programme for sustainable development aims to go beyond the traditional 'command and control' approach to stimulate more ecologically sound practices in private, public and not-for-profit sector organisations (Biondi and Frey 1995). In pursuing this objective, the Fifth Action programme according to Sunkin et al (1998, p.27):

- Focuses upon the agents and activities which deplete natural resources and damage the environment, rather than waiting for environmental problems to arise; and,
- Initiates changes in industrial trends and practices which are detrimental to the natural environment.

Overall, this change in philosophy represents a significant shift in policy substance of the four previous environmental action programmes, in response to the continued deterioration of the EU's environment as detailed in its *State of the Environment* report (Hillary 1995). Given this more progressive outlook the Fifth EAP proposes broadening the range of environmental policy instruments to complement traditional 'command and control' legislation. The Fifth EAP considers the most effective means of persuading enterprises to adopt a more proactive approach to environmental management to be voluntary and 'market-based'. The primary aim of such instruments are to introduce competition between industrial organisations on the basis of their environmental performance standards (Hillary 1995; Biondi and Frey 1995). The recent EMAS Regulation is one of the most innovative of such instruments. This initiative focuses upon industrial company sites and aims to publish information on environmental performance standards and give public recognition to enterprises that make a significant commitment to improve their environmental performance standards.

One of the main aims of the Fifth EAP is to create a 'virtuous circle' of environmental protection, conservation and improvement of the natural environment that includes companies, regulators and the general public. The EMAS Regulation is one environmental instrument designed to deliver this aim which lies at the heart of the Fifth EAP. The importance of the EMAS

Regulation in EU environmental policy is further reflected in the fact that the scheme is an Environmental Regulation, one of the stricter legal instruments available to European decision-makers. Furthermore, an EU Regulation is a legal instrument which takes precedence over national environmental laws and one which Member States must comply with.

European Industry Pollution and EMAS Implementation

The EMAS Regulation is principally directed towards Small and Medium-Sized Enterprises (SMEs) which comprise more than 99 per cent of the 15.7 million business organisations registered in the EU. SMEs have a significant impact upon Europe's natural environment and recent estimates suggest that this sector collectively contributes 70 per cent of all industrial pollution, with the other 30 per cent arising from the activities of large organisations (Rowe and Hollingsworth 1996, p.97). Having recognised this, the EMAS Regulation became publicly available for voluntary participation by large and small manufacturing organisations in 1995.

By February 1998 over forty industrial organisations had received the EMAS Regulation certificate. Thus far most of the EMAS Regulation implementing organisations have been very large companies and not the SMEs as anticipated by the EU. This is probably explained by the general lack of environmental consciousness found by many environmental surveys examining European SMEs (see, for example, Rowe and Hollingsworth 1996; Hodgson 1995; Vaughan and Mickle 1993; Barrow and Burnett 1990). However, the companies who have implemented the EMAS Regulation are important for the future viability of the initiative. This is because they are among the leading European industrial organisations and are the principal innovators in environmental management. Generally speaking, these are the organisations which other companies benchmark themselves against. It is possible to speculate that they will follow the lead of the implementing EMAS Regulation companies. Furthermore, as a result of their participation in the EMAS Regulation many companies are now pushing environmental issues down their organisational supply chain. The EMAS Regulation companies are now asking that their suppliers and contractors consider the natural environment in their strategic and operational decision-making more comprehensively than they have done previously. Should these organisations fail to respond to their requests then they may find themselves de-listed from future contracts. As a result of these developments it ought to become increasingly difficult to do business in Europe without having in place an EMS equivalent to ISO 14001 or the EMAS Regulation.

Components of the EMAS Regulation

Like ISO 14001, the EMAS Regulation requires organisations to formulate a comprehensive environmental management system (EMS). This should cover the range of the organisation's strategic and operational activities including administrative systems and human resource, quality, information and marketing functions. The purpose of the EMS established by the EMAS Regulation is both multi-faceted and complex. However, it can be inferred from reading the EMAS Regulation formal documentation (CEC 1993) that the aims of the EMAS EMS are twofold. First, the EMS aims to ensure that a firm achieves the goals and targets that it has set for itself through the establishment of an Environmental Assurance (E.A.) system. Second, it seeks to augment these targets with new ones as the firm learns more about the effects which its activities have on the environment. To meet these aims the EMAS Regulation consists of a number of components shown as a schematic model in Figure 12.1.

The main research instruments used during the course of the study were a postal questionnaire and telephone interviews, primarily because the participating companies were geographically dispersed throughout the UK and alternative research instruments would have been difficult to employ given constraints in the research teams resources.

Each EU Member State is required to appoint a 'Competent Body' to oversee the implementation of the EMAS Regulation. The Department of Environment (DoE) is the Government Department with responsibility in the UK. The DoE provided a list of all organisations which have received the EMAS Regulation award. Since this list of companies was small in number, the entire population of the EMAS Regulation certified companies was surveyed during the latter part of 1996 and early 1997. This included eighteen different sites of which two-thirds were large organisations and one-third were SMEs. The DoE also provided the names and contact addresses of personnel in each of the organisations including senior environmental, health, safety, quality and operational executives who were directly responsible for implementing the EMAS Regulation. One of the problems in undertaking a postal survey is identifying the correct person to send the questionnaire to and because the DoE solved this problem it helped to generate a response rate of 82 per cent. This is a high response rate compared to the Bris and Kok (1995) EMAS Regulation survey which generated a response of 12.5 per cent.

Figure 12.1 Main components of the EMAS

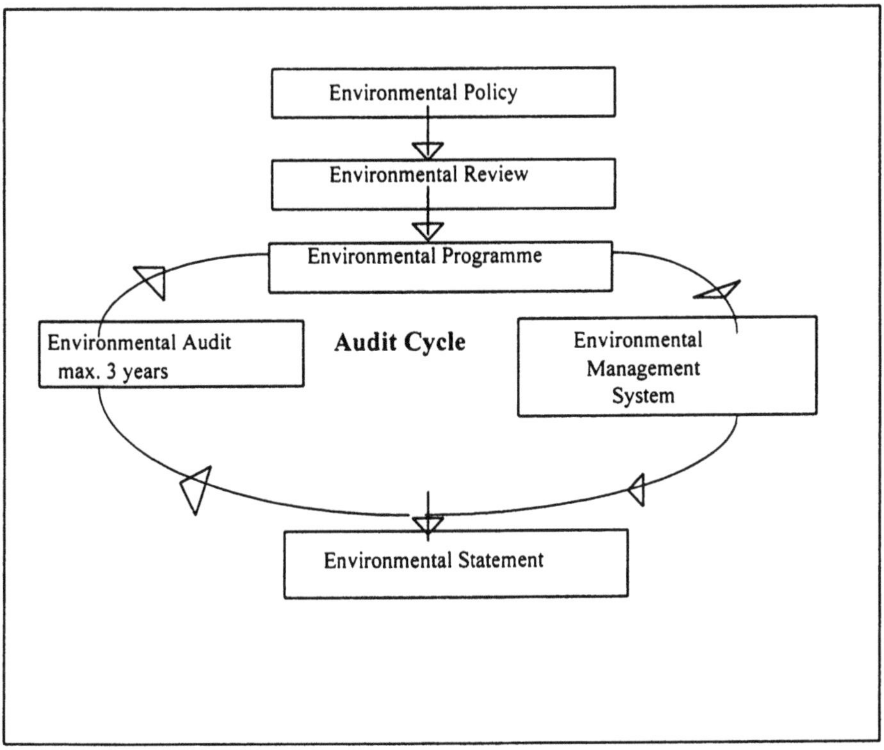

Key to Figure 12.1

1. The Preparatory Environmental Review. This is a review of the organisational site to identify all of its environmental 'effects' and this should include all of the organisation's inputs, throughputs and outputs. These inputs, throughputs and outputs should include, for example, the amount of energy that the organisation uses, its choice of raw materials, the types of wastes generated and the way in which the organisation's products are used and disposed of.

2. The Environmental Policy. This is a declaration of the organisation's aims and principles with regard to environmental issues. Four elements must be in place. First, a commitment to comply with EU and UK legislation. Second, a commitment to continuous environmental improvements. Third, the policy must be in a form which can be easily communicated to both internal and external stakeholders. Finally, the policy must also address relevant environmental issues which affect the organisation's activities and operations.

3. The Environmental Management Programme. The purpose of the environmental programme is to ensure that specific goals and targets are set in accordance with the environmental policy. Once environmental priorities are set, the programme then must be implemented into the organisation's management structures. The EMAS Regulation requires the organisation to establish clearly defined roles and responsibilities. The programme must also be supported by a clear administrative hierarchy and a manager must also be appointed to oversee the programme.

4. The Environmental Management System. This must be fully documented and integrated into the management structure and systems of the organisation. The administrative system must be supported by a management manual and this document needs to specify managers and staff responsibilities and also the formal interaction between all personnel. Written instructions must also be produced for all managers and staff.

5. The Environmental Audit. The purpose of the audit is to ensure that the organisation has conformed to the goals and targets set out in the management programme. The audit is to be carried out at periodic intervals and adopts the ICC definition of environmental auditing.

6. The Environmental Statement. A unique feature of the EMAS Regulation is that is requires a public statement linked to the environmental audit is to be published annually. The statement needs to be validated and accredited by an independent verifier and has to be made publicly available. This is to provide any interested parties with the opportunity to examine the environmental performance standards of the organisation and, for this reason, there is a requirement that the statement should be written in a user-friendly way.

The development of the postal and telephone schedules followed a comprehensive review of other EMS formal documentation (ISO 1995; CSA 1994 BSI 1994; NSAI 1994), the environmental guidelines and codes of practice articulated by leading industry bodies such as the International Chamber of Commerce (ICC) Business Charter for Sustainable Development (ICC 1991, 1994), the Coalition of Environmentally Responsible Economies (CERES) Principles (CERES 1995) and other research studies (Elliot et al, 1996; Peattie and Ringler 1995; Vaughan and Mickle 1993; Polonsky and Zeffane 1992; Zeffane and Polonsky 1995).

The research indicators focused upon the main components of the EMAS Regulation discussed above: the preparatory environmental review; the environmental policy; the environmental management programme; the environmental management systems; the environmental audit; and, the environmental statement. The indicators were grouped into three categories: the reasons given by the respondents for their participation in the EMAS Regulation; implementation issues; and the benefits resulting from participation in the EMAS Regulation.

Reasons for Participating in EMAS Regulation

The organisations surveyed were attracted to the scheme for various reasons and these responses are grouped under four headings, shown in descending order of importance in Table 12.1. These findings are generally supported by other researchers who have studied the benefits of participating in EMS including the EMAS Regulation (see, for example, Gebler *et al* 1997; Hillary 1997 and 1995; Riglar 1997; Biondi and Frey 1995; Bris and Kok 1995; Counsell and King 1995). In this study the highest ranked reason for participating in the EMAS Regulation was that it would improve their capacity to communicate the organisation's environmental performance standards and ongoing improvements to a wide range of internal and external company stakeholders (see also Gebler *et al* 1997, DoE 1996, 1995; Institute of Environmental Management 1995). One Safety, Health and Environment manger said that participation in the EMAS would help his organisation:

> ...in the short term to establish a reputation of an environmental champion to company stakeholders with all the benefits associated with such a role..and at the same time contribute to a more sustainable future for employees.

The same manager listed these internal and external company stakeholders as 'managers and other employees, customers, suppliers, regulators, and the local community'.

The second most common reason was that participation would *formalise* the organisation's approach to environmental management. In summary, the respondents stated that participation in the EMAS Regulation would provide a more systematic approach to their environmental management system and with third party verification they could demonstrate a proactive approach to company environmental stakeholders (see also, Institute of Environmental Management 1995; Bris and Kok 1995, Christensen and Nielsen 1995). Another Health, Safety and Environmental Manager said that:

> Participation in the EMAS will enable our company to establish a comprehensive environmental policy, set management objectives and targets, formulate a wide ranging environmental management programme and to raise environmental standards on a continual basis....This would be achieved through establishing a formalised and externally certified environmental system covering all of the organisation's activities and operations including planning, manufacturing, administrative systems and marketing functions'.

The third most common reason was that participation would allow their organisation to secure competitive advantage in their business sector (see also, Mudie 1995). This would take the form of improved market share, enhanced product quality and reduced operating costs. One of the environmental managers indicated that 'participation in the EMAS would allow their companies to more aggressively promote their firm as a 'green company' and reap the full commercial benefits that such an image can bring.' Another environmental manager replied that: 'participation would enable their company to reduce operating costs by more effectively identifying wasteful manufacturing processes.'

The fourth reason was that EMAS participation would help to ensure compliance with EU and UK legislative and regulatory regimes (see also, Gale 1995; Martinson 1995). All of the respondents recognised that environmental legislation from both sources would become more stringent in the future:

> Compliance with existing and future environmental legislation is arguably one of the most important strategic and operational challenges facing our organisation...and it is important that we help to shape the legislative and

> regulatory agenda by being proactive...our participation in the EMAS demonstrates our willingness to voluntarily raise the environmental performance of our organisation to regulators and at the same time contribute to the sustainable development agenda.

However, these benefits were not secured without the organisations experiencing some implementation 'hotspots' and it is to these that the analysis now turns.

Table 12.1 Reasons given for participating in the EMAS

Ranking (1= most important, 4 least important)	Reason/Motivation
1	Environmental Assurance to internal and external parties
2	Normalisation of the management system
3	Competitive Advantage
4	Compliance with EU and UK legislation and regulatory requirements

The Implementation 'Hotspots'

The terminology

Each of the respondents found the terminology of the formal EMAS Regulation documentation confusing and misleading. Of some concern were the terms 'environmental effects', 'review' and 'audit'. In this context the following statement made by one of the quality managers summarises the general view on this matter:

> We were very unsure what constituted the difference between an environmental review and the environmental audit...During the review we often confused these terms...When we consulted the EMAS requirements, it uses the two terms almost interchangeably.

All of the respondents stated that it is crucial to the future of the EMAS Regulation that these differences in terminology are clarified in the formal documentation. This problem in the use of terminology is not uncharacteristic of the environmental management literature (see, for example, Moxen and Strachan 1998; Ledgerwood 1997) or the EMAS literature which has examined the implementation of this scheme in other EU Member States (for further details, see, for example, Spencer-Cook 1997; Bris and Kok 1995; Biondi and Frey 1995; Counsell and King 1995).

The environmental review and audit

Following the preparatory environmental review and subsequent audits, some of the respondents experienced difficulties in setting environmental priorities (see Table 12.3) primarily because their organisations found it difficult to make enough sense to make use of the complex technical data generated by the review and audit process. The general research evidence presented by Spencer-Cook (1997), Biondi and Frey (1995) and Bris and Kok (1995) also support this finding. According to one quality manager:

> The environmental review and audits generated unlimited data on energy usage, water consumption, waste materials from manufacturing processes, unwanted by-products, effluents, emissions to air and solid wastes but we were unsure where our priorities should lie given all this data....Overall, we were very, very confused by all this data!

Financial costs

Many of the respondents also expressed concern over the cost of establishing and maintaining the EMAS Regulation (see Table 12.2 for the full descriptive statistics). Unfortunately, there are no other publicly available studies to compare with these figures (recently quoted in a United Nations Trade and Environment Expert Panel Meeting Report, 1997). The survey found that the initial cost of meeting the EMAS Regulation requirements ranged from £25,000 to £296,000 with an average cost of £94,600. In the case of ongoing costs, the study found that this ranged from a low of £3,000 to a high of £25,000 per annum with an average of £10,433 per annum. One senior manager explained that:

> These costs may act as a significant barrier to many other SMEs implementing the scheme into their operations. Many of these

> organisations may not be able to call upon the resources which we have at our own disposal.

This finding is supported by the academic literature. For example, an analysis by Spencer-Cook (1997) concludes by calling into question the financial viability of European SMEs implementing the EMAS Regulation. The major costs included: managerial time, the replacement and upgrading of existing manufacturing equipment, the establishment of new administrative systems and the appointment of new personnel.

Table 12.2 The costs associated with the EMAS regulation

Type of Statistics	Initial Cost	Ongoing cost per annum
Mean	£94,600	£10,433
Standard Deviation	£101,753	£10,559
Range	£271,000	£22,000
Maximum	£296,000	£25,000
Minimum	£25,000	£3,000

A bureaucratic system?

Many of the respondents stated that their organisations paid excessive attention to the management manual which was established to oversee the administrative system (for further details, see, for example, Moxen and Strachan 1998 and 1995; Strachan 1997ab; Biondi and Frey 1995; Bris and Kok 1995; Counsell and King 1995). To some this proved to be a 'bureaucratic minefield' and 'resulted in the overly elaborate documentation of highly specific goals and targets, including the specification of timing, resources and individual responsibilities, work instructions, purchasing policies, legal requirements and performance indicators'. While this was not a bad thing in itself, one of the respondents noted that little attention was given to the second goal of the EMAS Regulation, that of continually raising the environmental performance of the organisation in the longer term. This is an issue often referred to in the general environmental management literature and one which causes much concern to many EMS commentators (see, for example, Moxen and Strachan 1998; Faure *et al* 1993; Gilbert 1993).

Technical barriers to change

Table 12.3 ranks the main technical problems experienced by organisations participating in the EMAS Regulation in order of importance. As previously indicated, many of the respondents found it difficult to set priorities following the preparatory environmental review and subsequent audits. Many of those surveyed also found that there was a lack of technical skills available 'in-house' in these areas. This made the measurement and collection of environmental data generally difficult. In order to secure the necessary level of expertise some had to employ external consultants. This often this generated extra costs on top of implementing the EMAS Regulation. The lack of technical skills and the cost of employing consultants may act as a significant barrier to SMEs intending to implement EMAS (for further details, see, for example, Moxen and Strachan 1998 and 1995; Strachan 1997ab; Biondi and Frey 1995; Bris and Kok 1995; Post and Altman 1994).

Table 12.3 Types of technical problem encountered

Ranking (1=most significant, 4=least significant	Technical barriers to change
1	Availability of environmental data
2	Setting environmental priorities
3	Data collection
4	Lack of technical skills
5	Measurement and collection of environmental data
6	Lack of technical support from industry and government bodies

Organisational Barriers to Change

Table 12.4 ranks the main barriers to organisational change. Many of the respondents stated that managers were resistant to change and did not fully appreciate the need to implement the EMAS Regulation. Resistance to change was strongest in middle management, a finding which is characteristic of other studies which have examined implementation barriers during an environmental change programme (for further details, see, for example, Moxen and Strachan 1998 and 1995; Strachan 1997ab; Biondi and Frey 11995; Bris and Kok 1995; Counsell and King 1995). One quality manager, for example, said that this was because 'his people felt uncertain of what was expected of them' or that they 'did not want the inconvenience of adapting to change'. Another said that his staff felt:

> This was yet another change initiative that if they quietly resisted would disappear after a few months in operation. They had already seen quality management systems and business process re-engineering initiatives over recent years.

Many of the participating organisations added to this problem of resistance to change by failing to formulate effective training programmes for their managers and staff. One quality manager explained:

> We appear to have learned little from our early experiences of quality management...When BS 5750 (ISO 9001) was initially introduced.... training issues were neglected. This was despite the fact this would prove to be a critical success factor in the long term success of the system...We have forgotten this very quickly...This failure will be the downfall of many environmental management systems.

Another manager said that:

> Training is fundamental to the EMAS implementation process and that the human resource department has a very important role to play in this process...and without the knowledge of the human resource practitioner in managing the change process the EMAS will fail in securing its aims.

Finally, the respondents said that employee attitude, poor communication and past administrative systems were further structural and cultural barriers to change in the organisation.

Table 12.4 Organisational barriers to change

Ranking (1=most important, 3=least important	Organisational barriers to change
1	The bureaucracy attached to establishing the EMAS Regulation
2	Resistance to change by middle managers
3	Ineffective training programmes
4	Employee Attitude
5	Poor communication
6	Past practice and administrative systems

The Environmental Statement

Respondents generally found difficulty in producing the environmental statement. This arose from the requirement that the statement needed to appeal to the divergent and often conflicting agendas of many company stakeholders. Some also felt that the production of an environmental statement would expose their organisations to unnecessary criticism. However, most felt that the environmental statement was an important component of their marketing strategy and one which the organisation could use to promote their environmental performance standards to the outside world. Thus:

> Environmental communication is playing an ever increasing part in an organisations strategy. The sustainability challenge requires clear and verifiable information, words and facts rather than rhetoric - for the EMAS companies, commitment and performance are demonstrated through the publication of the environmental statement.

Additionally the publication of the environmental statement allowed the organisation to 'communicate their corporate culture, distinguish their vision, values and social responsibility and to seek effective dialogue to get their message across to a growing and critical audience'. (For further details of the benefits of green reporting, see, for example, Brink et al 1998).

The Benefits of EMAS Participation

Despite the implementation 'hotspots', a number of benefits have arisen as a result of having implemented the EMAS Regulation. The survey findings are ranked according to the importance of the benefits claimed by the respondents (summarised in Table 12.5). These findings are generally supported by previous research on EMS (see, for example, Gebler *et al* 1997; Hillary 1997 and 1995; Biondi and Frey 1995; Bris and Kok 1995; Counsell and King 1995).

The most important benefit was improved environmental assurance (EA). The respondents claimed that participation in the EMAS had allowed their organisation to establish an environmental management system that enabled them to meet and exceed environmental targets and to demonstrate this to key organisational stakeholders (see also, Bris and Kok 1995; Institute of Environmental Management 1995). As one environmental manager said:

> The most important benefit that arose from our participation in the EMAS Regulation was that it allowed us to establish a system of management that enabled us to set goals and targets and to meet and exceed those targets. Also it allowed us to demonstrate this to parties external to the formal organisation.

The second ranked benefit was securing competitive advantage, through increased market share (though, interestingly, none of the organisations were able to quantify this yet) and reduced operating costs from efficiency savings in raw materials, energy efficiency, the manufacturing process, material recovery and packaging (see also DoE 1995, 1996; Christensen and Nielsen 1995; Mudie 1995).

The third most common benefit was to comply with EU and UK legislation. The respondents claimed that as a result of their participation in the EMAS Regulation they were better able to identify and meet the demands of regulatory agencies. Though ranked lowest, some respondents claimed 'this would help to guarantee the future commercial viability of the enterprise' (see also, Gale 1995; Martinson 1995).

Table 12.5 The benefits of participation in the EMAS

Ranking (1= most important, 3 least important)	Benefits
1	Environmental Assurance to internal and external parties
2	Competitive Advantage
3	Compliance with EU and UK legislation and regulatory requirements

Conclusion and Recommendations

The EMAS Regulation is the most recent approach to raising environmental performance standards in European organisations. The scheme requires organisations to establish and implement environmental policies, programmes and management systems, on a site-by-site basis; to systematically audit environmental performance standards; and to provide independently validated information on their environmental performance standards to internal and external company stakeholders through the publication of an environmental statement (Bris and Kok 1995). The EMAS Regulation goes beyond 'command and control' legislation by recognising that the management systems of an organisation have an important part to play in raising environmental performance standards across the range of its activities.

Many UK organisations are now implementing the scheme. This study found that these organisations were attracted to the initiative for numerous reasons including the prospect of more effective environmental audit, the normalisation of the organisation's approach to managing the natural environment, the anticipation of competitive advantage and more effective compliance with EU and UK legislative/regulatory regimes. While many of these opportunities were claimed to have been realised in practice, some difficulties - or 'implementation hotspots' - were experienced: confusing and misleading terminology; difficulties in setting environmental priorities; financial costs; bureaucracy; the lack of technical skills; resistance to change; and the production of an environmental statement.

These implementation hotspots support the findings made by other management and environmental researchers. Post and Altman (1994), for example, have identified a range of industry and organisational barriers when implementing an EMS. The industry barriers are concerned with the specific environmental problems and issues facing the enterprise, whereas the organisational barriers are not unique to environmental issues but may affect the implementation of any type of strategic or operational management change programme. The industry barriers outlined by Post and Altman include: competitive rivalry, industry regulation, community concern, the configuration of current operating practices, technical knowledge and information gaps. The organisational barriers include: the attitude of managers and staff, communication, past practice and the attitude of middle and top management and cost. Such organisational barriers are further outlined by Moxen and Strachan (1995) and Strachan (1997ab).

Given that EMAS Regulation and ISO 14001 implementation is still in its infancy, it is hoped that the findings generated by this survey will be of assistance to practitioners, educationalists and management consultants and to organisations participating in EMAS Regulation in the future, in order that they secure the full societal and organisational benefits of the scheme. While this study has only examined the situation in the UK, further research should be undertaken on the situation in other EU states to assess the EMAS Regulation implementation process elsewhere. This is all the more important given that, in 1998, the European Commission embarks on the consultative review process on the operation of the EMAS as mandated in the Regulation and this is likely to see the scheme extended to other industrial sectors. It is hoped that this initial study has provided a useful starting point for this review process.

The successful implementation of the EMAS Regulation and other EMS such as ISO 14001 necessitates action by the business community, government and other decision-makers. In addition, measures including technical assistance and capacity-building are needed to assist companies in setting up EMS. The chapter concludes with the following recommendations for the Scottish Parliament in the effective promotion of business EMS.

1. Training and Awareness Raising in EMS Implementation. Training and awareness raising efforts are required to demonstrate the need and the potential benefits of EMS in Scotland. These efforts could target CBI Scotland and leading business leaders, Scottish Enterprise and the Local Enterprise Company (LEC) network. In the short term training and awareness efforts are important, particularly with companies exporting to

Europe and beyond. An examination of national implementation experiences as well as the identification of international training packages would also be useful for future participating organisations in EMS such as the EMAS Regulation and ISO 14001.

2. The Dissemination of Environmental Information and the Formation of Best Practice Programmes. Information on trends in the use of EMS is essential, in particular for industrial sectors where EMS are most likely to become an important factor in the market place (e.g. manufacturing industries) or are most likely to benefit from EMS implementation because of their significant impact upon the natural environment (e.g. oil and gas and construction industries). Best practice and technical assistance programmes could identify such information, and organisational networking opportunities may also be usefully applied in this context.

3. Pilot Schemes. The available evidence suggests that pilot schemes are an effective means to improve understanding of EMS implementation within and between organisations. Such activities also offer learning opportunities for the business community, government, certification bodies and other actors. A cross-national exchange of experience could also prove fruitful.

4. Industry Cooperation and Government Grant Schemes. While larger organisations have initially lead the way in securing the EMAS Regulation, SMEs also need encouragement to apply. Recognising that SMEs may need help in carrying out the various implementation steps the grant scheme linked to the EMAS Regulation (Small Company Environmental and Energy Management Assistance Scheme, SCEEMAS) may need to be revised and extended to meet the needs of smaller and less well equipped organisations more effectively.

5. Enhancing Scottish Participation in the Growing Market for Environmental Services. The environmental industry is one of the fastest growing European industrial sectors and Scottish organisations need to find ways to promote participation in the growing market for services associated with the wider use of EMS including consultants, training and certification. Government decision-makers have a key role to play in creating the infrastructure to support this.

References

Barrow, C. and Burnett (1990), *How Green Are Small Companies?* Cranfield School of Management.

Biondi, V. and Frey, M. (1995), 'Participation in the EU Eco-Management and Audit Scheme: An Analysis of Small-and Medium Sized Enterprises', *European Environment*, 5, 5, pp. 128-133.

Brink, P.T., Haines, R., Owen, S., Smith, D., and Whitaker, B. (1998), 'Consulting the Stakeholder: A New Approach to Environmental Reporting for IBM (UK) Ltd' in Ledgerwood, G., Greening the Boardroom: Corporate Governance and Business Sustainability, Greenleaf Publishing, Sheffield.

Bris, H.S. and Kok, EHJ (1995), 'The Eco-Management and Audit Scheme: Experience and Benefits', Eco-Management and Auditing, Vol. 2, pp. 80-84.

BSI (1994) Specifications for Environmental Management System BS 7750, BSI, London.

Canadian Standards Association (CSA) (1994) *A Voluntary Environmental Management System: Z750-94, CSA*, Ontario, Canada.

CEC (1992) *The EC 5th Environmental Action Programme*, Official Journal of the European Communities, Brussels.

CEC (1993) *Council Regulation (EEC) No1836/93 of 29 June 1993 amended proposal for a council regulation (EEC) allowing voluntary participation by companies in the industrial sector in a Community eco-management and audit scheme*, Official Journal of the European Communities, Brussels.

CERES (1995) *Guide to the CERES Principles*, Coalition for Environmentally Responsible Economies, Boston.

Christensen, P. and Nielsen, E.H. (1995) 'Implementing Environmental Management Systems in Danish Industry: Do we go Beyond Compliance?', *The Annual Business Strategy and the Environment Conference Proceedings*, ERP Environment, Shipley.

Counsell, D., and King, B. (1995) 'Implementation of Environmental Policy Using the Eco-Management and Audit Scheme for UK Local Government: Experience in Cleveland', *European Environment*, 5, 4, pp. 91-97.

DoE (1996) *EMAS: A Catalyst for Change*, DoE, London.

DoE (1995) *EC Eco-management and Auditing Scheme: A Participant's Guide*, DoE, London.

Elliot, D., Patton, D., and Lenagham, C., (1996) 'UK Business and Environmental Strategy: A Survey and Analysis of East Midland Firms Approaches to Environmental Audit', *Greener Management International*, 13, pp.30-48.

Faure, M., Vervaele, J., and Weale, A., (1993) *Environmental Standards in the European Union in an Interdisciplinary Framework*, Metro, Antwerp.

Franke, J. and Watzold, F., (1995) 'Political Evolution of EMAS: Perspectives from the EU, National Governments and Industrial Groups', *European Environment*, 5, 6, pp.155-159.

Gabel, H.L. and Sinclair-Desgange, B. (1993), 'Managerial Incentives and Environmental Compliance'. *Journal of Environmental Economics and Management*, Vol. 24, No. 2, pp. 229-240.

Gale, R.J.P (1995, 'The Canadian Standards Association's Approach to Environmental Auditing', *Eco-Management and Auditing*, Vol. 2, pp. 36-46.

Gelber, M., Hanf, B. and Huther, S. (1997), 'EMAS Implementation at Hipp in Germany', in Sheldon, C., *ISO 14001 and Beyond: Environmental Management Systems in the Real World*, Greenleaf Publishing, Sheffield.

Gilbert, M.J. (1993). *Achieving Environmental Management Standards: A Step-by-Step Guide to Meeting BS 7750*, Pitman Publishing, London.

Hillary, R. (1997), 'Environmental Management Standards: What do SMEs Think?, in Sheldon, C., Sheldon, C., *ISO 14001 and Beyond: Environmental Management Systems in the Real World*, Greenleaf Publishing, Sheffield.

Hillary, R (1995) 'The Eco-Management and Auditing Scheme as a vehicle to promote cleaner production', *Eco-Management and Auditing*, 2, pp.1-7.

Hodgson, S.B. (1995), SMEs and Environmental Management. The European Experience. *Eco-Management and Auditing*, Vol. 2, pp. 85-89.

ICC (1994) *Principles for Environmental Management - Summary Report of a Seminar Organised by ICC United Kingdom*, ICC, London.

ICC Working Party on Sustainable Development (1991) *Background Note on the ICC 'Business Charter For Sustainable Development' Doc 210/364*, ICC, Paris.

Institute of Environmental Management (1995) 'Eco-management and Audit Scheme: enhancing the impact of environmental management', *Journal of the Institute of Environmental Management*, 3, 3.

ISO (1995) *Committee Draft ISO/CD, Environmental Management Systems*, Geneva, Switzerland.

Ledgerwood, G. (1998), *Greening the Boardroom: Corporate Governance and Business Sustainability*, Greenleaf Publishing, Sheffield.

Martinson, J. (1995) 'Accuracy Put Before Cost'. *Financial Times*, 2 June, 1995.

Marthijsen, P.S.R.F (1995). *A Guide to European Union Law*, 6th ed, Sweet & Maxwell, London.

Moxen, J. and Strachan, P. (1998), *Managing Green Teams: Environmental Change in Organisations and Networks*, Greenleaf Publishing, Sheffield.

Moxen, J., and Strachan, P. (1995), 'The Formulation of Standards for Environmental Management Systems: Structural and Cultural Issues', *Greener Management International*, 12, pp.32-48.

Mudie, P. (1995), 'Profiting From Improved Environmental Performance', *Eco-Management and Auditing*, Vol 2, pp.71-75.

NSAI (1994), *IS 310: 1994 Environmental Management Systems - Guiding Principles and Requirements*, National Standards Authority of Ireland, Dublin.

Peattie, K. and Ringler, A., (1995), 'Responding to the Green Challenge: A Manufacturing/Service Sector Comparison', *Business Strategy and the Environment Conference*, ERP Environment, Shipley.

Polonsky, M., and Zeffane, R., (1992), 'Corporate Environmental Commitment in Australia: A Sectorial Comparison', *Business Strategy and the Environment*, 1, 2, 25-39.

Post, J.E. and Altman, B. (1994), 'Managing the Environmental Change Process: Barriers and Opportunities', *Journal of Organisational Change Management*, 7, 4, pp.65-81.

Riglar, N. (1997), 'Eco-Management and Audit Scheme for UK Local Authorities: Three Years On', in Sheldon, C., *ISO 14001 and Beyond: Environmental Management Systems in the Real World*, Greenleaf Publishing, Sheffield.

Rowe, J., and Hollingsworth, D. (1996), 'Improving the Environmental Performance of Small and Medium Sized Enterprises: A Study in Avon'. *Eco-Management and Auditing*, Vol. 3 No. 2, pp. 97-107.

Schmidheiny, S. (1992), *Changing Course: A Global Business Perspective on Development and the Environment*, MIT Press, London.

Spencer-Cook, A. (1997), 'From EMAS to SMAS: Charting the Course From Environmental Management to Sustainability', in Sheldon, C., *ISO 14001 and Beyond: Environmental Management Systems in the Real World*, Greenleaf Publishing, Sheffield.

Strachan, P. (1996) 'Managing Change and the Environment: The Recommendations of a Business Workshop', *Eco-Management and Auditing*, 3, 2, pp.69-75.

Strachan, P. (1997a) 'Should Environmental Management Standards be a Mechanistic Control System or a Framework for Learning', *The Learning Organisation*, Vol. 4, No 1, pp.10-17.

Strachan, P., Haque, M., McCulloch, A., and Moxen, J. (1997b) 'The Eco-Management and Audit Scheme: Recent Experiences of UK Participating Organisations', *European Environment*, Vol 7, 1, pp.25-33.

Sunkin, M., Ong, M.G. and Wright, R. (1998), *Sourcebook on Environmental Law*, Cavendish Publishing, London.

United Nations (UN) Conference on Trade and Development (1997) 'Expert Meeting on Trade and Investment Impacts of Environmental Management Standards, Particularly the ISO 14001 Series on Developing Countries', 29-31 October 1997, UN, Geneva.

Vaughan, D., and Mickle, C (1993) *Environmental Profiles of European Business*, Earthscan, London.

Zeffane, R., and Polonsky, M., (1995) 'Corporate Environmental Commitment: Developing the Operational Concept', *Business Strategy and the Environment*, 3,4, 1995.

Wolf, S. and White, A. (1995) *Environmental Law*, Cavendish Publishing, London.

13 Sustainable Development in a Small Country: The global and European agenda

KEVIN DUNION

Introduction

The Scottish Parliament has been given the challenging, indeed daunting responsibility for guiding us towards sustainable development. Some of the powers necessary for the job are being devolved to Edinburgh including environmental protection policy and matters relating to air, land and water pollution; the natural built heritage; and water supplies and sewerage. Agriculture, fisheries and forestry, as well as economic development and transport, will also be within the Parliament's responsibility. However truly integrated activity on domestic matters is limited because control of railways and energy for example will remain with Westminster.

There are many real improvements in these areas which can be made to contribute towards a more sustainable future for Scotland. But the reality is that, however essential it is to further improve the quality of Scotland's sewerage emissions or urban air quality, these are as Lord Sewel has said only 'frontline skirmishes of a much larger battle'. Whilst we may have the captains of industry and politics now taking up commissions on our own doorstep, the generals and field marshals of this global campaign are still directing affairs from London, Brussels, Paris and New York. It can be argued that the Scottish Parliament is being given responsibility at the very moment when states, far less sub-states, have a markedly diminished ability to implement a distinctive or vanguard set of policies. The Scottish Parliament will be subject to the forces and policies emanating from international and European institutions. The scope to influence these is intentionally limited. Furthermore large institutions such as the European Union and the UN are setting the framework for equally powerful players, the trans national corporations, to pursue an agenda which is even more removed from the possibility of Scotland's citizens and its Parliament having any influence.

Political and financial tensions

We have to ask whether, with the responsibility for sustainable development, there will be a willingness at Holyrood to take a leading role in pushing the boundaries of action towards environmental protection and sustainability. Those who have argued for years in favour of a Scottish Parliament have hoped and imagined that it would have the powers to escape from the straitjacket of policies which were either inappropriate for Scottish circumstances or apparently disregarded Scottish opinion - a state of affairs which was particularly pronounced in the 1980s. At that time an alternative view was expressed which built on the rejection of free market values and presumed that traditional Scottish values of democracy, collective responsibility and recognition of the community as well as the individual would be at the forefront of future policy making. We can still hope that this will be embodied in the new Parliament.

However we also have to allow for the possibility that there may be a reluctance and indeed a resistance to implementing environmental and sustainable development policies emanating from international and European institutions. This may be particularly so where the Parliament perceives potential adverse consequences for Scotland in terms of lost jobs and increased costs from measures to protect the environment. Of course the resistance is likely to be even more marked if it is the case that the international or EU agreement has been negotiated by a UK administration which has a strained relationship with the Scottish Parliament or indeed an entirely different political complexion.

In addition to political strains, there will be financial concerns. The Scottish Parliament has a limited capacity to alter its financial base and its tax varying powers are restricted to the basic rate of income tax. This may turn out to be an ill thought-out and unnecessarily narrow financial settlement. Increasingly we can expect the burden of taxation in the UK, as throughout Europe, to shift from taxes on labour, to taxes on pollution and resources. This may make it difficult for the Scottish Parliament to actually use its income tax raising powers and yet it has limited capacity to use these indirect taxes. Furthermore the indirect taxes may have a perceived differential impact upon those who have limited capacity to respond to such price signals.

We have already seen examples of this with the fuel duty escalator, which generates a backlash at each budget because it is felt that those in more sparsely populated areas in Scotland are being asked to pay the price of what they see as largely an urban problem. (Over 145,000 people in

Scotland live more than two hours drive from the nearest service centre, ie a town of more than 30,000 people.)

Tensions are also likely to arise over the environmental responsibilities inherited by the Scottish Parliament for Sites of Special Scientific Interest (SSSIs), Ramsar sites, National Scenic Areas and others, if alterations to the Barnett formula fail to take into account the land mass of Scotland and the conservation and ecological responsibilities that this brings to the Parliament.

International Pressures

Internationally the dominant issue which the Scottish Parliament has to comes to terms with is the globalisation of markets. The key players in this are the World Trade Organisation and the Organisation of Economic Corporation and Development. WTO was set up as a successor to the General Agreement on Tariffs and Trade (GATT). The WTO was determined to stamp out protectionism and trade restrictions. It is opposed to northern countries using concerns over for example the environmental costs of the production of materials or indeed the labour conditions under which they are produced, as the reason to erect trade barriers against a product or a country. The WTO Director General Renato Ruggiero has said . the risk now is of an insidious neo- protectionism which could try to use trade restrictions as a response to widespread concerns over labour, social or environmental standards'. The most widely celebrated case has been the challenge to the US Government's decision to prohibit the import of tuna from Mexico on the grounds that the numbers of dolphin killed when catching the tuna exceeded the limits set on US domestic fishing fleets. This has given rise to concern that other trade restrictions imposed for sound environmental reasons will also be challenged by the World Trade Organisation.

However so far the GATT and WTO have never ruled that trade related environment measures contained in an international agreement contravenes GATT. For example the Montreal Protocol (which phases out certain ozone depleting chemicals in trade between OECD parties and at the same time bans import of such products from non parties) has been an extremely successful international agreement. As many as 136 countries representing 99 per cent of the world's population, including all major producers of ozone depleting substances, have signed up to the Treaty. It is widely held that it is the ability to impose trade restrictions on non treaty parties which have significantly encouraged wide participation and compliance. Yet the prospect of a future clash is now said to have a 'chilling effect' on national and international environmental law-making, discouraging the use of what is called 'command and control' regulations and economic instruments which may fall foul of WTO rules.

The WTO has established a Committee on trade and the environment to try and resolve the conflict of, on the one hand, the political push towards globalisation with unrestricted markets, and on the other, the legitimate use of trade restrictions on those who flout environmental standards. However after three years the Committee has failed to reach agreement on any substantive issue and groups like Friends of the Earth have called for it to be replaced by an Intergovernmental Panel on Trade, Environment and Sustainability which would involve a wider constituency of players including a balance of environment, development and trade representatives of governments.

The WTO has a dominant position in respect of sustainability, firstly from a technical perspective in having the capacity to impose fines upon countries which it holds to have placed improper restrictions in respect of a specific trade (such as the dolphin/tuna case) but also because it is setting a political agenda which transforms the structure of commerce and the development of nations. It is joined in this by the OECD which is seeking to do for investment what the WTO is doing for trade. The hotly disputed Multilateral Agreement on Investments (MAI) is the centrepiece of its plans.

The purpose of this draft agreement would be to lubricate international investment flows and to prohibit national restrictions which operated to the disadvantage of foreign investment. Global capital flows totalled around $350 billion dollars in 1996 and the prospect of this free flow of capital alarms both those who see the lack of a safeguard for the environment or labour standards and a loss of national sovereignty. Governments in the North including the UK dismiss the clamour of concerns. Clare Short, Secretary of State for International Development, speaking in Edinburgh March 1998 was scornful of the 'Armageddon Now' tendency of environmental groups and said that Britain would not sign the MAI unless there were environmental and labour clauses within it. OECD negotiators also dismiss criticisms saying 'this isn't some dark conspiracy to undermine the basis of civilisation. Like the North American Free Trade Agreement, the MAI has become a scapegoat for popular anxiety about globalisation'.

Critics however point to the operation of NAFTA which has seen the US Ethyl Corporation sue the Canadian Government for having the temerity to outlaw the use of its additive MMT on the grounds that it is a toxic pollutant. They also point to the US waste disposal companies, who are suing Mexican local authorities, for refusing to allow plans for toxic landfill sites and designating the areas for nature protection instead, as the future under the MAI.

As with the WTO there is a perception that environmental and labour standards would always be optional extras to be disposed of in favour of maintaining free trade and globalisation. Critics point to the attempts by the International Chamber of Commerce to block the inclusion of measures to safeguard the environment and labour standards in the MAI as 'excess baggage'. Furthermore they contrast the non binding nature of the clauses on environmental, labour and social issues with the binding investor state dispute mechanisms which uphold companies' economic rights in the Treaty. The MAI therefore threatens national environmental and social laws.

These agreements will have their impact upon Scotland's economy and will limit the scope for action by its Parliament. Land reform provides a good example. There are many in the Highlands, including Jim Hunter, who have argued that if the Scottish Parliament is not able to address the issue of land reform it is hardly worth having at all. Concern has been expressed over the inappropriate use of the land, its use as an asset to be bought and sold, and problems over absentee landlordism. Many have cast envious glances at Scandinavian countries including Denmark where there are restrictions on land ownership which relate to the use of the land and whether the owner intends to live on it. Hugh Raven pursues these issues in Chapter 11 of this volume.

Under the MAI it would be almost impossible for Scotland to introduce reforms which embodied these characteristics, as these could be seen to unfairly prejudice the ability of a foreign investor, compared to a Scottish investor. By contrast the Danish Government has in its contribution to the MAI negotiations listed a series of exemptions to the operation of MAI in its country and one of those is to safeguard its current land legislation.

The drive to globalisation now dominates the world agenda. Transnational corporations are key players in developing the kind of political context in which they would like to operate. This is tacit acknowledgement of their scale. Fifty one of the world's largest economies are corporations. The combined sales of the world's top 200 corporations are equal to 28 per cent of the world's GDP. These 200 corporations employ only 0.33 per cent of the world's people.

Given the scale of such enterprises, globalisation does not only mean homogenising the market place for big business, but also regarding big business as an institution in its own right. The High Level Advisory Group on the Environment's Report to the OECD Secretary General in 1997 warned:

> Governments are losing much of their traditional influence on the policies that deeply concern their people such as equity, public health, social benefits for the elderly and the poor, environmental quality and taxation. Democratic elections, the backbone of the western ideal of freedom, are thereby losing much of their significance.

This is a profoundly depressing message for those who look forward to the establishment of a democratic institution in Edinburgh as a lever to tackle just such concerns. There is no doubt that increasingly transnational corporations are becoming key players in shaping the kind of political context in which they wish to operate. For instance at the Earth Summit in New York in 1997, the Secretary General of the UN organised a lunch with the chief executive officers of 10 TNCs to chart a formalisation of corporate involvement in the affairs of the UN. (It is interesting to note that the US Government sent Larry Summers, Deputy Secretary of the Treasury to represent its interests, who as former chief economist of the World Bank advocated the shipping of more toxic waste to low income countries because they are under polluted).

Given such powerful forces, it can be said that the Scottish Parliament will be in no worse position than any other. But my feeling, nevertheless, is that it is going to give rise to new frustration within Scotland. Firstly we have an aspiration of a Scotland which is more at ease with itself. People's hopes for the future seem to harness nostalgia for progress. A sustainable Scotland would include within it citizens doing a useful job in recognisably Scottish companies which had a concern for the community in which they were based, using more local resources in a sustainable manner and retaining more of the value of that work in Scotland. There will be a greater emphasis on building a domestic market, in supporting indigenous businesses and exploring how local economies can be revived with the benefits of new technology and a shift away from manufacturing to service provision. Instead of making cars and washing machines, we would have regional service centres which upgraded, repaired and dismantled modern machines, which were originally designed with resource efficiency in mind. Or local shops which also acted as collection points for customers who have ordered goods by computer from regional supermarkets.

It is not impossible for these aspirations to come about but they may be frustrated by countervailing pressures coming from global corporations who do not share such an agenda, and who have been quoted by economic development agencies who wish to remove environmental, social and local planning impediments to investment by increasingly footloose businesses.

The Scottish Parliament's input into these global discussions and decisions will be limited or non-existent. These are not matters which generally relate to devolved concerns. In particular it has been made clear that foreign affairs, including the ability to conclude international agreements in both reserved and devolved areas, should be the sole preserve of Westminster. But given the scale of the impact on what may be the preferred Scottish agenda, it is unlikely that Scottish Parliamentarians will not express an opinion. The devolution White Paper has made it clear that any issue can be discussed by the Scottish Parliament and furthermore that Scottish interests can be represented on UK delegation to international bodies.

Whatever the limitations, the Parliament should maximise the opportunities which do exist. It is interesting to note for example that the UK delegation to the UN General Assembly Special Session in June 1997 included not just the Prime Minister, Deputy Prime Minister, Foreign Secretary and Minister for International Development but also representatives from the Local Government Management Board, the Eco-labelling Board, the Environment Agency (for England and Wales) and ICI. In addition provision was made for representation from Guernsey and from Gibraltar. There is absolutely no reason then why the Scottish Parliament should not be represented on this and similar Government delegations.(For example the Chief Executive of Scottish Natural Heritage attended the World Conservation Congress in Montreal in 1996 as part of the official UK delegation). Such scope for practical engagement is admittedly limited and the willingness to take advantage of it is likely to be patchy depending upon what catches the political mood in Scotland at the time. However we should not ignore the opportunities which do exist.

European Legislation

When negotiations take place at the level of the European Union, it seems certain that the Scottish Parliament, notwithstanding what the founding legislation says, will demand that its voice is heard. Environmental legislation and other measures towards sustainability will be dominated by decisions taken within the European Union. The Amsterdam Summit of 1997 made sustainable development an explicit EU objective. European leaders also highlighted the requirement that environmental protection be integrated into the definition and implementation of EU policies and activities. Thus environmental protection requirements must be integrated into each of the EU's areas of activity: agriculture, transport, trade, regional policy and so on. Unlike most discussions at the UN or in the OECD, the direct consequences of decisions taken within the European Union will be felt within Scotland, often at a very local level.

The Scottish Parliament directly, and through its regulatory agencies such as Scottish Environment Protection Agency (SEPA), will have the responsibility for implementing EU obligations which concern devolved matters. The legislation makes it clear that, so far as devolved matters are concerned, the Scottish Parliament can either adopt UK legislation or pass its own laws. However it says 'where EU obligations are to be implemented separately for Scotland, there will be arrangements with the UK Government to ensure that the differences of approach are compatible for the need for consistency of effect'. Given the way in which EU legislation is framed, the *actual* room for manoeuvre may be more limited than separate implementation suggests. It has been argued that the influence of EU environmental legislation on national policy is playing an increasingly important role in reducing the flexibility and discretion which have traditionally characterised the regulatory process. This is because EU legislation is generally more detailed than national legislation in terms of the provisions contained within its instruments. For example the Nitrates Directive contains particular classifications by which to determine whether a waterway should be categorised as polluted.

This will have a significant impact upon the Scottish Parliament's function. Traditionally Scottish regulatory agencies have had an element of discretion to take into account local circumstances when controlling pollution. It has to be said that this has in the past also given rise to frustration on the part of environmental organisations who felt that this was too often used to permit poor standards of compliance with environmental laws. It has also created a culture of only meeting the minimum standards required by law and putting off investment to meet such standards to the last minute. Consequently the Parliament is going to inherit a legacy of under-investment in environmental protection which may cause it to fall foul of EC Directives. Many Scottish beaches, particularly in Ayrshire and the Forth and Clyde Estuaries, currently fail to meet the minimum standards of the EC Bathing Water directive, even though the date for compliance has now passed. Despite expensive upgrading of sewerage systems by the Water Authorities in the West of Scotland, the failures continue. The situation is likely to get even worse if the Scottish Office review of designated bathing beaches increases the number of officially monitored beaches from the current 23 to the 60-100 locations which environmental groups are calling for. An end to the dumping of sewage sludge at sea will also mean expensive alternative disposal routes. Future costs can be anticipated, such as a draft EC Directive requiring separate collection of hazardous waste from households and other municipal sources including paints, solvents, aerosols and fluorescent tubes.

It is quite clear that failure to comply with the EC Directives will result in an action being taken against the Scottish Parliament. The cost of any fine is levied by the European Commission will be borne entirely by the Scottish Parliament. More dramatically the UK Parliament may seek to impose a solution upon Scotland, even in areas which have been devolved to the Scottish Parliament, on the grounds that it is constitutionally responsible for meeting the undertakings which it is has given in international obligations. Certainly we should not be sanguine about avoiding meeting EC obligations. The Commission has started legal action against thirteen states for failing either to adopt or to communicate their plans for toxic waste disposal. A Commission spokesman is quoted as saying 'past action by the Commission indicated that the threat of financial penalties could up compliance remarkably'.

Scotland is likely to want to have a say on those matters which are devolved and, more generally, on the issue of sustainable development within the European Union. This represents something of a problem for Westminster as the proposals within the founding legislation demonstrate. The Devolution White Paper (Scottish Office, 1997) says:

> The guiding principle is that the UK should be able to speak with one voice in the international arena and to advance policies (for example in international negotiations) which take proper account of the interests of all parts of the UK. It will also be essential that the UK Government is in a position to implement obligations it has undertaken in good faith internationally or which are imposed on the UK by international law.

However so far as the EU is concerned the document says 'the existence of clear and distinctive domestic voices from regional government is already a source of strength for other member states'. It is proposed therefore that the Scottish Executive's ministers and officials should be fully involved in discussions within the UK Government about the formulation of the UK's policy position on all issues which touch on devolved matters. However the White Paper sternly warns 'this will require of course mutual respect for the confidentiality of those discussions and adherence to the resultant UK line, without which it would be impossible to maintain such close working relationships'.

It is difficult to see how this is going to work in practice. It is scarcely imaginable that, on a controversial issue, the Scottish Minister responsible would be unable to report to the Scottish Parliament on the progress of discussions with his or her Westminster counterpart. Secondly, adherence to the resultant UK line suggests that in negotiations within the EU itself the UK line will be maintained without alteration but of course flexibility in negotiation is essential, especially where majority voting is concerned. In this case Holyrood's representative may feel that Scottish interests are adversely affected by the compromise being negotiated, and have failed to be taken into account.

Matters are likely to be exacerbated of course where the Westminster representative is of a different political persuasion from the Scottish representative either because of a change of government in the UK as a whole or because of coalition government in Scotland. Second guessing these eventualities is important and has a bearing even upon how Scotland is represented in Brussels. If a harmonious relationship is presumed then it seems logical to be an attached if distinctive part of the UK representative office. That would provide detailed insight into current negotiations and a regular and direct input into the UK position. However if there is an expectation of forming alliances and exerting influence behind the scenes with a recognisably Scottish voice then the temptation may be to clearly delineate the Scottish presence by, for example, locating the Scottish Parliamentary Brussels office alongside Scotland Europa in Square de Meus.

Much will depend on the stance which we expect the Scottish Parliament to take towards Europe. Clearly there is concern that Scotland will be dragging its feet on implementing EU Directives and obstructing future tightening of European environmental laws and sustainable development policies on the grounds of the perceived higher costs or lost jobs. Initiatives such as the Carbon Energy Tax, which is being supported by Commissioner Ritt Bjerregaard, or a European Air Fuel Taxation Levy are likely to produce a negative response in Scotland. Such measures are meant to prompt a change in behaviour not simply exact fiscal revenues. Yet the Scottish Parliament has not been equipped with the necessary powers to respond to such signals. For example it has limited powers over rail regulation to offset air fuel levies which are intended to discourage short haul flights and it is constrained by Treasury Public Sector borrowing requirements on its ability to invest in energy efficiency measures to offset a carbon energy tax. If Scotland is to avoid a reputation for simply whingeing about such matters or being environmentally reactionary, it is essential that we have a robust ability to represent Scottish concerns at an early stage in formulating the UK position and to secure methods to mitigate negative impacts where this is deserved because of social geography, without denying the sound sustainability principles which have brought forth such measures.

It is also important to avoid the impression that the EU is some kind of milch cow. The Agenda 2000 expansion plans of the EU have prompted headlines like 'Highlands lose millions in new Euro Masterplan' or less luridly 'European Funding for Scotland under threat'. Approximately 85 per cent of Scotland's population is covered by the various programmes which have provided more than £1.6 billion pounds since 1979. Clearly the changes in structure funds and in agricultural subsidies through the CAP reforms will have a significant impact in Scotland, but an approach based simply upon maintaining the flow of cash as opposed to contributing to the sustainable development of an enlarged Europe would deservedly be seen as parochial. In particular we need a rather more honest assessment of just how well the structure funds have been used in Scotland to assist in targeting the funds to accession countries such as Poland, the Czech Republic, Hungary, Slovenia and Estonia in a fashion which both genuinely assists their transition and contributes towards sustainable development.

It is not unreasonable to consider what should be done in the event that a Scottish Parliament *should* act in a parochial manner towards international/EU obligations or seek to frustrate their implementation. But we should not be so pessimistic. The purpose of securing the Parliament is to ensure that policies are better fitted to Scotland and better reflect Scottish opinion based on a presumption that the collective will is progressive and outward-looking. The political settlement is likely to produce a more pluralistic parliament bolstered by an extensive and internationally engaged civil society of NGOs, trade unions and the churches.

We might equally well consider what would happen if Scotland was to take the lead on some of these issues. What if the Scottish Parliament sought to implement EC Directives more quickly than Westminster or more extensively? Scottish Parliamentarians would want a guarantee for example that, should they do so and Westminster dragged its feet and was eventually fined by the European Commission, then no part of this charge would be borne by Scotland.

It is also likely that notwithstanding the strictures in the White Paper section on International Relations, the Scottish Parliament will form strong international bonds and will make its views known outside of formal treaty negotiations. Indeed this may help to modernise the UK position. The idea of a univocal voice within international negotiations is a myth but one which the UK still cleaves to. For a distinctively Scottish voice to be heard within the EU and its institutions, and indeed in other international gatherings, will mean the Scottish Parliament augmenting its professional staff with those who have experience of such negotiations. Furthermore such expertise should not be used simply to prepare Scotland for implementing whatever the UK has negotiated but also to prepare the UK for responding to initiatives coming from within Scotland.

Conclusion

There are strong reasons why Scotland should adopt high environmental standards not least because of the comparative advantage in attracting progressive high value investment and securing a well qualified workforce. It is well accepted that a programme of energy efficiency would generate jobs, improve health and quality of life and reduce CO_2 emissions. The need for such a programme is more urgent and compelling than anywhere else in the UK. Anticipating and acting to secure higher air quality or better bathing water, and implementing even more stringent planning regulations or even higher standards of building regulations, would of course mean adopting differential standards north and south of the Border. The capacity to do so is the whole point of devolution.

Index

For Product Safety Concerns and Information please contact our EU
representative GPSR@taylorandfrancis.com
Taylor & Francis Verlag GmbH, Kaufingerstraße 24, 80331 München, Germany

www.ingramcontent.com/pod-product-compliance
Ingram Content Group UK Ltd.
Pitfield, Milton Keynes, MK11 3LW, UK
UKHW022203190926
475702UK00003B/172

* 9 7 8 1 1 3 8 3 1 3 1 2 5 *